MATERIALS RESEARCH SOCIETY
SYMPOSIUM PROCEEDINGS VOLUME **1666**

Film-Silicon Science and Technology

April 21-25, 2014
San Francisco, California, USA

Printed from e-media with permission by:

Curran Associates, Inc.
57 Morehouse Lane
Red Hook, NY 12571
www.proceedings.com

ISBN: 978-1-5108-0527-9

Some format issues inherent in the e-media version may also appear in this print version.

©Materials Research Society 2014

This reprint is produced with the permission of the Materials
Research Society and Cambridge University Press.

This publication is in copyright, subject to statutory exception and to the
provisions of relevant collective licensing agreements. No reproduction
of any part may take place without the written permission of Cambridge
University Press.

Cambridge University Press
Cambridge, New York, Melbourne, Madrid, Cape Town,
Singapore, São Paulo, Delhi, Tokyo, Mexico City

Cambridge University Press
32 Avenue of the Americas, New York, NY 10013-2473, USA
www.cambridge.org

Materials Research Society
506 Keystone Drive, Warrendale, PA 15086
www.mrs.org

CODEN: MRSPDH

ISBN: 978-1-5108-0527-9

Cambridge University Press has no responsibility for the persistence or
accuracy of URLs for external or third-part Internet Web sites referred to
in this publication and does not guarantee that any content on such Web sites
is, or will remain, accurate or appropriate.

Additional copies of this publication are available from:

Curran Associates, Inc.
57 Morehouse Lane
Red Hook, NY 12571 USA
Phone: 845-758-0400
Fax: 845-758-2634
Email: curran@proceedings.com
Web: www.proceedings.com

Film-Silicon Science and Technology

Materials Research Society Symposium Proceedings
Volume 1666

San Francisco, California, USA
21-25 April 2014

TABLE OF CONTENTS

Pulsed–light Crystallization of Thin Film Silicon, Germanium, and Silicon Germanium Alloy 1
B. Yan, W. Toner, M. Dubey, Q. Fan, C.-S. Jiang, D. Stevenson

Improved Metastability and Performance of Amorphous Silicon Solar Cells 7
T. Matsui, A. Bidiville, H. Sai, T. Suezaki, M. Matsumoto, K. Saito, I. Yoshida, M. Kondo

Minority Carrier Annihilation at Crystalline Silicon Interface in Metal Oxide Semiconductor Structure 18
J. Furukawa, S. Shigeno, S. Yoshidomi, T. Node, M. Hasumi, T. Sameshima, T. Mizuno

Direct Gap Group IV Semiconductors for Next Generation Si-based IR Photonics 24
J. Kouvetakis, J. Gallagher, J. Menendez

Effect of Annealing on Microstructure in (Doped and Undoped) Hydrogenated Amorphous Silicon Films 36
W. Beyer, W. Hilgers, D. Lennartz, F. Maier, N. Nickel, F. Pennartz, P. Prunici

Hydrogenated Amorphous Silicon Germanium by Hot Wire CVD As an Alternative for Microcrystalline Silicon in Tandem and Triple Junction Solar Cells 42
L. Veldhuizen, Y. Kuang, N. Bakker, C. Werf, S. Yun, R. Schropp

High Quality Kerfless Silicon Mono-crystalline Wafers and Cells by High Throughput Epitaxial Growth 48
R. Hao, T. Ravi, V. Siva, J. Vatus, D. Miller, J. Custodio, K. Moyers

Optimization of the Protocrystalline p-layer in a-Si:h-based n-i-p Photodiodes 59
Y. Vygranenko, M. Fernandes, M. Vieira, A. Sazonov

Near-UV Background As a Bridge Between Visible and Infrared Communication 65
M. Vieira, V. Silva, I. Rodrigues, P. Louro

Increased Sensitivity in a-SiC Pinpin Multilayers in the VIS-NIR Range Under UV Light 71
V. Silva, I. Rodrigues, M. Vieira, P. Louro

Optical Characterization of Si Nanowires: Dependence with Substrate Orientation and Light Polarization 78
J. Badan, R. Marotti, E. Dalchiele, D. Ariosa, F. Martin, D. Leinen, J. Ramos-Barrado

Femtosecond Laser Materials Processing of a-Si:H Below the Ablation Threshold 85
B. Soleymanzadeh, W. Beyer, F. Luekermann, P. Prunici, W. Pfeiffer, H. Stiebig

Crystallization of Amorphous Silicon and Dopant Activation using Xenon Flash-Lamp Annealing (FLA) 91
T. Mudgal, C. Reepmeyer, R. Manley, D. Cormier, K. Hirschman

Defects and Doping in Nanocrystalline Silicon-Germanium Devices 97
S. Konduri, W. Mulder, V. Dalal

Tunable and Wireless Photoimpedance Light Sensor 103
T. Saxena, S. Rumyantsev, P. Dutta, M. Shur

Defects in Epitaxial Lift-off Thin Si Films/Wafers and Their Influence on the Solar Cell Performance 109
B. Sopori, S. Devayajanam, P. Basnyat, H. Moutinho, R. Reedy, K. VanSant, T. Ravi, R. Hao, J. Vatus, S. Nag

Crystallization of Amorphous Silicon Thin Films by Microwave Heating 115
T. Nakamura, S. Yoshidomi, M. Hasumi, T. Sameshima, T. Mizuno

Author Index

Mater. Res. Soc. Symp. Proc. Vol. 1666 © 2014 Materials Research Society
DOI: 10.1557/opl.2014.668

Pulsed–light Crystallization of Thin Film Silicon, Germanium, and Silicon Germanium Alloy

Baojie Yan[1], William Toner[1], Mukul Dubey[2], Qihua Fan[2], Chun-Sheng Jiang[3], David Stevenson[1]

[1.] Wintek Electro-Optics Corporation, Ann Arbor, Michigan

[2.] Department of Electrical Engineering and Computer Science, South Dakota State University, Brookings, South Dakoda

[3.] National Renewable Energy Laboratory, Golden, Colorado

ABSTRACT

We report the recent progress of crystallization of amorphous silicon (a-Si), amorphous germanium (a-Ge) and amorphous silicon germanium alloy (a-SiGe) using a pulsed-Xenon-lamp system with multiple lamps. The precursor materials were deposited using a sputtering machine on display glass substrates maintained on a rotary holder. The RF powers on the silicon and germanium targets were varied to control the Ge/Si ratio in the materials. The film thickness was in the range of 50-100 nm, targeting the application in thin film transistors (TFT). The samples were pre-heated to 350-450°C in a conveyer chamber with nitrogen flow before the crystallization. The materials were characterized using AFM, Raman and Spectroscopic Ellipsometry. We demonstrated that we can uniformly crystallize a-Si, a-SiGe, and a-Ge with a single-pulse or multiple-pulse process on 10×5 cm^2 glass substrates. We found that the required crystallization power for a-Ge is much lower than for a-Si. The power needed to crystallize a-SiGe is between the power required for a-Ge and a-Si crystallizations, and it increased with increasing Si fraction. No Raman signal was measurable in the as-deposited films. Strong Raman peaks at 520 cm^{-1} and 290 cm^{-1} were observed in the pulsed-lamp crystallized poly-Si and poly-Ge films, respectively. Distinct Ge-Ge, Si-Ge, and Si-Si vibration modes were observed at ~285 cm^{-1}, ~390 cm^{-1}, and ~470 cm^{-1}, respectively, in the poly-SiGe films formed after the pulsed-light treatments. Their intensity ratios and the peak positions depended on the Ge/Si ratio and the light intensity used for the crystallization. AFM images showed the formation of large grains with increased surface roughness.

INTRODUCTION

Thin film silicon and silicon germanium alloy materials used in solar cells and thin film transistors (TFT) have evolved from hydrogenated amorphous silicon (a-Si:H) to hydrogenated nanocrystalline silicon (nc-Si:H) to large grain polycrystalline silicon (poly-Si). nc-Si:H films have provided significant improvements in the efficiency of thin film silicon solar cells [1,2]. However, thin film silicon solar modules still have lower efficiency than c-Si modules, and, as a result, struggle to compete in the market. One method for improving the efficiency of thin film Si solar cells is to make large grain poly-Si films [3,4]. Large-grain, low-temperature polycrystalline silicon thin films (LTPS) with high mobility are also needed for the TFTs used in high resolution LCD and OLED displays [5-8]. Currently the LTPS TFTs are made using laser-induced crystallization or metal-induced solid phase crystallization of a-Si:H. These are expensive methods that are difficult to scale to large size displays. A low-cost process for producing thin film poly-Si or possibly poly-SiGe is highly desirable for both photovoltaic and display applications.

Pulsed-lamp-induced crystallization of a-Si, a-Ge and a-SiGe has provided a low cost process for making poly-Si based materials [9-11]. Previous research focused mainly on thick materials for solar cell applications. For the channel layers in TFTs, the thickness is normally in the range of 50-100 nm. Because of the low total absorption in such thin layers, it is difficult to crystallize the thin films used in TFTs using the pulsed Xenon lamps.

In this paper, we have carried out systematic studies of pulsed-lamp crystallization of very thin a-Si, a-Ge, and a-SiGe alloy, deposited by a sputtering system. For the application in TFTs, we focused on the film thickness of 50-120 nm.

EXPERIMENTAL

We deposited a-Si, a-Ge, and a-SiGe alloy films on 0.5-mm thick display glass using an RF sputtering system, wherein the substrates were maintained on a rotary holder such that the substrate moved continuously through the Si plasma zone and the Ge plasma zone. The Ge/Si ratio was controlled by the RF powers on the Ge and Si targets. The sputtering deposition was performed in Ar at 35 mtorr with no hydrogen. Therefore, a de-hydrogenation process was not needed. The film thicknesses were measured using optical methods (Filmetrics and spectroscopic ellipsometry (SE)). The film thicknesses were in the range of 50-120 nm.

The pulsed-lamp crystallization was carried out using a multi-lamp system with three Xenon lamps in parallel (Novacentrix, PulseForge 3300), which can process a sample size of 10×5 cm^2 uniformly. A conveyer system with a pre-heating chamber was placed under the Xenon lamp array with a quartz window. The samples were heated to 350-450°C before the pulsed-lamp illumination. The length of the light pulse was varied from 150 μs to 400 μs. The material properties were characterized using Raman, AFM, and SE.

RESULTS AND DISCUSSION

Poly-Si films

We first carried out the crystallization of a-Si films. The left plot in Fig. 1 shows Raman spectra of two poly-Si films formed by the pulsed-light process. A c-Si wafer was used as the reference. The right plot is the spectra normalized to their peak height. The material structure changes depend on the pulsed-light intensity and duration, hence the energy delivered by the light pulse. The thin a-Si films can be crystallized by a single pulse. The process parameters for the two samples in Fig. 1 were 470 V for the driving voltage of the Xenon lamps for both

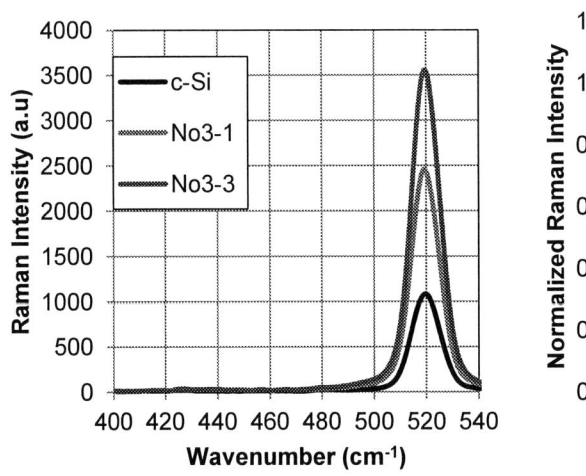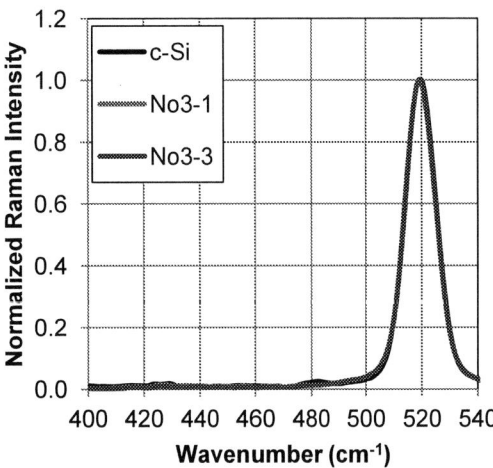

Figure 1. (left) as measured and (right) normalized Raman spectra of two poly-Si films and one c-Si wafer. The poly-Si samples were made with pulse-lamp crystallization of sputtering deposited a-Si films.

samples, and 400 µs and 300 µs pulse durations for samples No3-1 and No3-3, respectively. One can see that the a-Si films were completely changed into poly-Si. The line shape is the same as that of the c-Si wafer. One may notice that the line width of the reference c-Si spectrum is wider than the real c-Si spectrum. The lines are broadened by the Raman spectroscopy used in this study.

Figure 2 shows the AFM images of an as-deposited a-Si film and a poly-Si film crystallized by the pulsed-lamp process. Some very fine structures are evident on the a-Si film with a root-mean-square (RMS) roughness of 2.4 nm, whereas the poly-Si film has large features of size of up to 0.5 µm, and the RMS roughness was increased to 22 nm. We also did SE

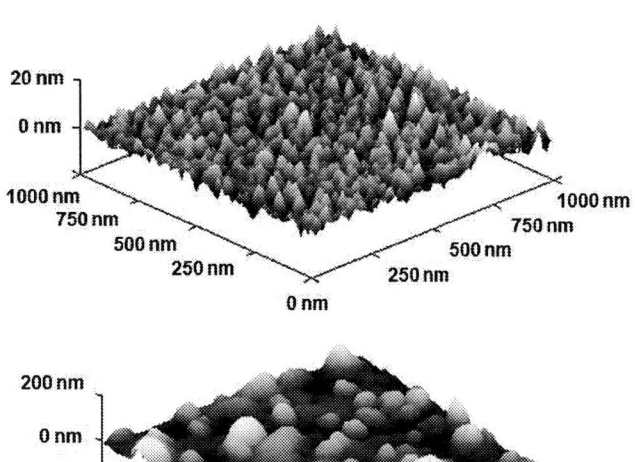

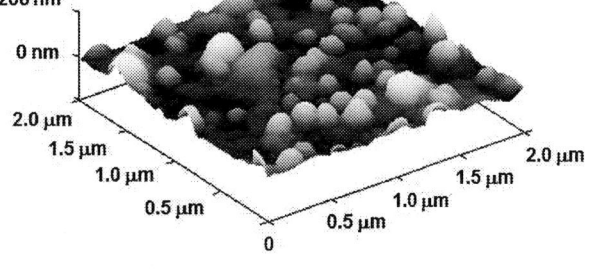

Figure 2. AFM images of (top) a precursor a-Si film and (bottom) a poly-Si film formed by the pulsed-lamp crystallization. Note that the top figure has a size of 1×1 µm^2 and vertical scale of 20 nm, but the bottom one is 2×2 µm^2 with a 200 nm vertical scale.

analysis on the poly-Si film and found a very good fit to a double-layer with 100% poly-Si in the bulk of the film and a 30-40 nm surface layer with an average 15% void density, which is consistent with the AFM images.

Poly-Ge

Ge has a much lower melting temperature than Si, and therefore needs a much lower energy for crystallization. We carried out the pulse-lamp crystallization on the sputtered a-Ge. Figure 3 plots Raman spectra of the as-deposited a-Ge film and the poly-Ge films formed by the pulsed-lamp crystallization. The crystallization was performed with a much lower light power than the process for making the poly-Si. The driving voltage for the Xenon lamps was 340 V, and the duration was 300 μs. Figure 4 shows an AFM image of a poly-Ge film. Compared to the poly-Si films, the poly-Ge films are very flat with an RMS roughness of 0.7 nm. The pulsed-lamp process does not increase the surface roughness when a proper low power is used, which could be important for TFT fabrication.

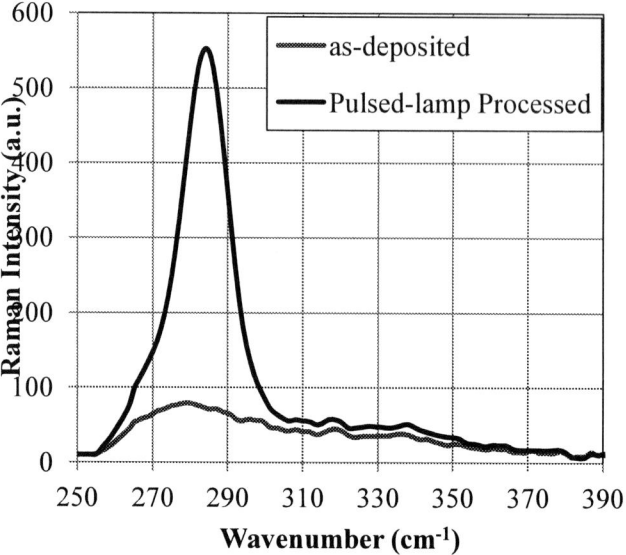

Figure 3. Raman spectra of a-Ge and poly-Ge formed by the pulsed-lamp crystallization.

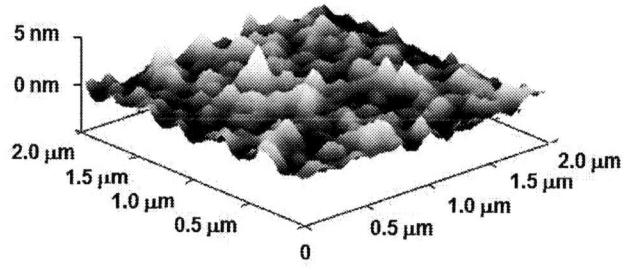

Figure 4. An AFM image of poly-Si films. The vertical scale is only 5 nm, and the RMS roughness is only 0.7 nm.

Poly-SiGe

Poly-SiGe could be a potential useful material for high mobility TFTs because of the lower process temperature and higher carrier mobility compared to poly-Si [12-13]. In this study, poly-SiGe films were also made using the pulsed-lamp process. Table I lists the Ge contents of three typical as-deposited a-SiGe films together with the pulsed-lamp process parameters. One can see that the energy in the light pulse is decreased with increasing Ge content. Figure 5 plots the Raman spectra of three poly-SiGe films. One may make the following observations. First, three distinguishable Raman peaks are observed with the Si-Si TO mode in the range of 470-500 cm^{-1}, the Si-Ge TO mode in the range of 370-400 cm^{-1}, and the Ge-Ge TO mode in the range of 250-300 cm^{-1}, indicating that the Si and Ge not only forms an alloy, but also forms separated crystalline Si and crystalline Ge phases. Second, the intensity ratio of the Ge-Ge peak to the Si-Si peak increases with the increasing Ge content. Third, the Si-Si peak position shifts to lower energy compared to the Si-Si peak in c-Si, and the peak shift increases with increasing Ge content. The Si-Si TO mode shifts to lower energy in poly-SiGe films have been observed previously and explained by the influence of the surrounding Ge atoms [14-15].

Similar to poly-Si, the pulsed-lamp crystallized poly-SiGe also showed a large increase in the surface roughness and feature sizes. The increased surface roughness could be a potential issue for TFT device fabrication, but it could be beneficial for enhanced light trapping for solar cell applications.

Table I: The Ge contents of three a-SiGe precursor films and the pulsed-lamp process conditions, where V is the voltage driving the lamp, T is the duration of the light pulse, and E is the energy density in the light pulse.

Sample No	Ge content	V (V)	T (μs)	E (mJ/cm^2)
A	33%	580	150	4641
B	19%	470	400	7055
C	66%	600	120	3848

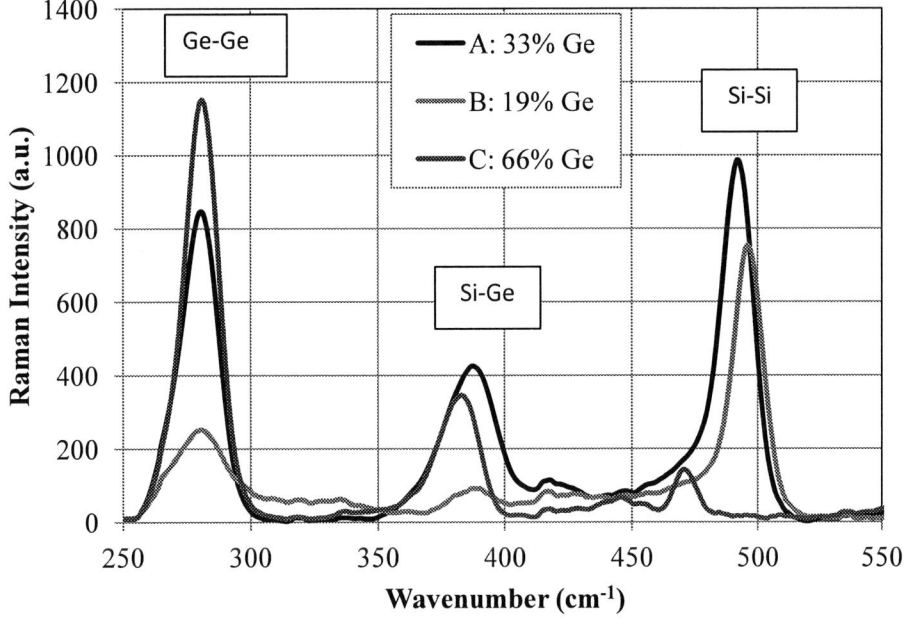

Figure 5. Raman spectra of poly-SiGe films formed by the pulsed-lamp process. The three samples were made with different Ge contents.

SUMMARY

We have carried out systematic experiments on the pulsed-lamp crystallization of a-Si, a-Ge, and a-SiGe films deposited by sputtering for TFT applications. Very thin films of 50-120 nm thickness can be fully crystallized to form poly-Si, poly-Ge and poly-SiGe films throughout a large area of 10×5 cm^2 with a single light pulse. The threshold energy (or power) for crystallizing a-Si films is much higher than for crystallizing a-Ge films; it decreases with increasing Ge content in a-SiGe films. The Raman spectra show that the pulsed-lamp processed materials are fully crystallized with no observable amorphous phase. Distinct Si-Si TO, Si-Ge TO, and Ge-Ge TO modes are observed in the poly-SiGe films. The Ge-Ge/Si-Si peak ratio increases with the increasing Ge content. The Si-Si peak position shifts to the low energy side due to the incorporation of Ge in the films. The poly-Si and poly-SiGe have a significantly increased surface roughness compared to the a-Si and a-SiGe precursors, which could be a potential issue for TFT fabrication, but it could benefit thin film solar cell performance because of the additional light trapping by the surface texture.

REFERENCES

1. Thin-Film Silicon Solar Cells, edited by A. Shah, EPFL Press, 2010, Lausanne, Switzerland.
2. S. Guha, J. Yang, and B.Yan, Solar Energy Materials & Solar Cells, **119**, 1 (2013).
3. J. Haschke, L. Jogschies, D. Amkreutz, L. Korte, and B. Rech, Solar Energy Materials & Solar Cells, **115**, 7 (2013).
4. S. Varlamova, J. Dore, R. Evans, et al., Solar Energy Materials & Solar Cells, **119**, 246 (2013).
5. L. Xu and C. P. Grigoropoulosa, J. Appl. Phys. **99**, 034508 (2006).
6. T. Pier, K. Kandoussi, C. Simon, N. Coulon, H. Lhermite, T. Mohammed-Brahim, J. F. Bergamini, Thin Solid Films **515**, 7585 (2007).
7. L. Maioloa, A. Pecoraa, F. Maitaa, et al., Sensors and Actuators B **179** 114 (2013).
8. C.-L. Wang, I-C. Lee, C.-Y. Wu, Y.-T. Cheng, P.-Y. Yang, H.-C. Cheng, Thin Solid Films **529**, 421 (2013).
9. K. Ohdaira, N. Tomura, S. Ishii, and H. Matsumura, Thin Solid Films **519**, 4459 (2011)
10. T. Nishikawa, K. Ohdaira, and H. Matsumura, Current Appl. Phys. **11** 604 (2011).
11. K. Ohdaira and H.Matsumura, J. Cryst. Growth, **362**, 149 (2013).
12. V. Subramanian and K. C. Saraswat, IEEE Tran. Electron Devices. **45**, 1690 (1998).
13. Z. Jin, H. S. Kwok, and M. Wong, IEEE Tran. Electron Devices Lett. **19**, 502 (1998).
14. T. S. Perova, J. Wasyluk, K. Lyutovich, E. Kasper, M. Oehme, K. Rode, and A. Waldron, J. Appl. Phys. **109**, 033502 (2011).
15. O. Pages, R. H. Hussein, and V. J. B. Torres, J. Appl. Phys. **114**, 033513 (2013).

Mater. Res. Soc. Symp. Proc. Vol. 1666 © 2014 Materials Research Society
DOI: 10.1557/opl.2014.918

Improved metastability and performance of amorphous silicon solar cells

Takuya Matsui[1], Adrien Bidiville[1], Hitoshi Sai[1], Takashi Suezaki[2,3], Mitsuhiro Matsumoto[2,4], Kimihiko Saito[2,5], Isao Yoshida[2] and Michio Kondo[1]

[1]National Institute of Advanced Industrial Science and Technology (AIST), 1-1-1 Umezono, Tsukuba, Ibaraki, 305-8568 Japan
[2]Photovoltaic Power Generation Technology Research Association (PVTEC), 1-1-1 Umezono, Tsukuba, Ibaraki, 305-8568 Japan
[3]Kaneka Corporation, 157-34 Kamiyoshidai, Toyooka, Hyougo, 668-0831 Japan
[4]Panasonic Corporation, 3-4 Hikaridai, Seika-cho, Soraku-gun, Kyoto, 619-0237 Japan
[5]Fukushima University, 1 Kanayagawa, Fukushima, Fukushima, 960-1296 Japan

ABSTRACT

We show that high-efficiency and low-degradation hydrogenated amorphous silicon (a-Si:H) p-i-n solar cells can be obtained by depositing absorber layers in a triode-type plasma-enhanced chemical vapor deposition (PECVD) process. Although the deposition rate is relatively low (0.01-0.03 nm/s) compared to the conventional diode-type PECVD process (~0.2 nm/s), the light-induced degradation in conversion efficiency of single-junction solar cell is substantially reduced ($\Delta\eta/\eta_{ini}$~10%) due to the suppression of light-induced metastable defects in the a-Si:H absorber layer. So far, we have attained an independently-confirmed stabilized efficiency of 10.11% for a 220-nm-thick a-Si:H solar cell which was light soaked under 1 sun illumination for 1000 hours at cell temperature of 50°C. We further demonstrate that stabilized efficiencies as high as 10% can be maintained even when the solar cell is thickened to >300 nm.

INTRODUCTION

Hydrogenated amorphous silicon (a-Si:H) and microcrystalline silicon (μc-Si:H) films grown by plasma-enhanced chemical vapor deposition (PECVD) are extensively employed as light absorbers in thin-film silicon solar cells. Although the high-efficiency (~14-15%) tandem devices have been demonstrated in the initial state [1,2], the stabilized efficiencies after prolonged illumination are limited to ~12% [3-7] due to the light-induced degradation of a-Si:H, known as the Staebler-Wronski effect [8]. Because of this adverse effect, the performance of thin-film silicon solar cells remains nearly half of that of high-efficiency crystalline silicon solar cells. So far, a variety of deposition techniques have been proposed to improve the light-soaking stability of a-Si:H. Nevertheless, the substantial reduction of the light-induced degradation has not been satisfactorily demonstrated, particularly for the cells exhibiting high initial efficiency.

In our previous studies [9,10], we have shown that high-efficiency and low-degradation a-Si:H solar cells can be obtained when the a-Si:H absorber layer is deposited by the triode

PECVD technique. Although the deposition rate is relatively low (0.01-0.03 nm/s) compared to the conventional diode-type PECVD process (~0.2 nm/s), the light-induced degradation in conversion efficiency ($\Delta\eta/\eta_{ini}$) of single-junction solar cell is substantially reduced. A stabilized efficiency of 9.6% has been achieved using a commercially-available TCO substrate [9].

In this contribution, we report on the further progress of a-Si:H single-junction solar cell development realized by optimizing not only the absorber layer quality but also other component layers. We show that stabilized efficiencies as high as 10% can be attained for a range of absorber layer thickness varying from less than 200 to more than 300 nm. Results of the material characterization such as microstructure parameters and light-induced metastable defects in the a-Si:H layers and devices are also discussed.

EXPERIMENT

Undoped a-Si:H layers were grown in a triode PECVD reactor in which a mesh electrode was placed between powered and grounded electrodes. A schematic diagram of the triode-type PECVD is shown in Fig. 1. The plasma was generated with a 60 MHz excitation at a power density of 60 mW/cm^2, a pressure of ~10 Pa and SiH$_4$/H$_2$ flow rates of 20/20 sccm. A mesh electrode was placed 20 mm above the powered electrode. A DC voltage of -25 V was applied to the mesh to confine the SiH$_4$-H$_2$ glow-discharge between the powered and mesh electrodes. As a result, no visible plasma exists between the mesh and grounded electrodes [10]. In our triode deposition setup, the deposition rate can be controlled in the range from ~0.01 to ~0.05 nm/s by adjusting the distance between the mesh and substrate (L_{M-S}) without changing plasma-generation condition [10]. As a reference material, a-Si:H films were also deposited by our standard diode PECVD system (13.56 MHz) at a deposition rate of 0.25 nm/s. The materials were characterized by transmission-reflection spectroscopy and Fourier-transform infrared

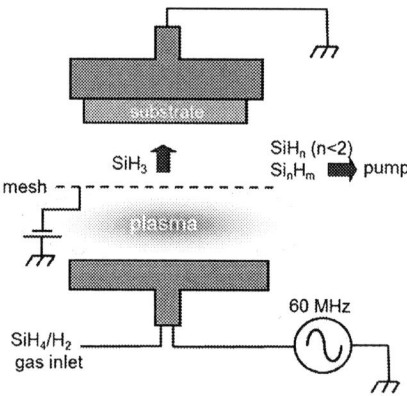

Fig. 1. A schematic diagram of a SiH$_4$-H$_2$ glow discharge in the triode PECVD reactor.

spectroscopy (FTIR) for measurements of the optical band gap and Si-H_n bond density, respectively. The evolution of metastable defects in a-Si:H was evaluated by measuring the sub band gap absorption by means of Fourier-transform photocurrent spectroscopy (FTPS) [11-13].

The intrinsic a-Si:H layers grown either by triode or diode PECVD were integrated into a p-i-n device in a superstrate configuration, as shown in Fig. 2. For the front transparent conductive oxide (TCO) substrate, SnO_2-coated glass provided by Asahi Glass Company (AGC) was used. To improve the light in-coupling at the TCO/Si interface, the SnO_2 surface was covered with antireflection layers consisting of TiO_2 (35 nm)-ZnO (10 nm) [14]. The 10 nm-thick ZnO layer acts as a protection layer from the hydrogen-induced reduction reaction of TiO_2. For the doped layers, we deposited p- and n-layers consisting of a μc-Si:H (p)/a-SiC:H (p)/a-Si:H (undoped) triple layer and μc-Si:H (n) single layer, respectively, in separate reactors. The device area (1.04 cm^2) was defined by a patterned back contact consisting of ZnO:Al/Ag/ZnO:Ga stacked layers deposited by sputtering. The ZnO:Al/Ag layers act as a back reflector, while the ZnO:Ga layer was used for protection of Ag in the post-deposition processes. For some solar cells, to reduce the contribution of the peripheral dark current component, an edge-isolation process was carried out by reactive plasma etching. Then, solar cells were annealed at 160°C in vacuum for 2 hours. The current density-voltage (J-V) characteristics of the solar cells were measured under standard air mass 1.5 (AM1.5) global illumination condition (irradiance: 100 mW/cm^2, cell temperature: T_{cell}=25 °C). To avoid overestimation of the photocurrent of the solar cells, we used a mask to define an illumination area (1 cm^2) that was slightly smaller than the device area (1.04 cm^2). The solar cells were exposed to AM1.5 illumination under 100 mW/cm^2 intensity (T_{cell}=50 °C, 1000 h, open circuit) [15]. For some selected samples, J-V characteristics after light soaking were measured by the Calibration, Standards and Measurement Team (CSMT) of AIST for high accuracy cell characterization.

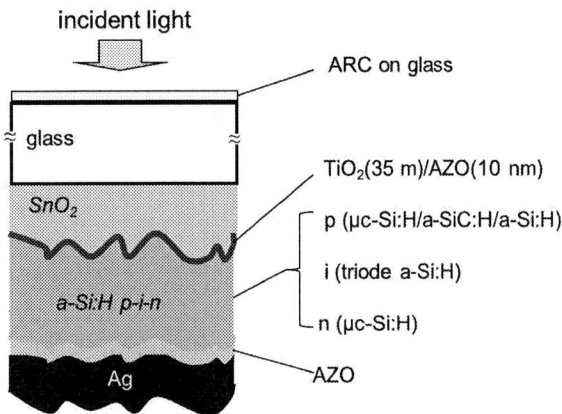

Fig. 2. Device structure of a-Si:H single-junction p-i-n solar cell.

RESULTS AND DISCUSSION

Deposition of stable a-Si:H films and solar cells

It is widely known that the light-soaking stability of a-Si:H films [16] and solar cells [17] is improved as the density of dihydoride (Si-H_2) bond (or density of Si-H_n (n=1-3) bonds in the internal void surface) decreases in a-Si:H. Furthermore, it has been argued that the Si-H_2 bond density can be reduced by preventing the incorporation of reactive species such as SiH_2 and higher silane radicals (Si$_n$H$_m$, n>2) into the film during the a-Si:H deposition [18]. For this purpose, the triode electrode configuration in the PECVD process has been proposed as a radical separation technique [19, 20], which provides stable a-Si:H films with less Si-H_2 bond density as compared to the films deposited by the conventional PECVD with a diode electrode configuration.

In the triode PECVD process, the growth of a-Si:H is dominated by the diffusion of long-lifetime radicals such as SiH_3 [18]. Thus, the incorporation of short-lifetime radicals such as SiH_2 and higher silane radicals can be reduced by moving the surface of the growing film away from the plasma generation zone. Although the deposition rate is reduced by an order of magnitude for triode PECVD, the material parameters such as the optical band gap and hydrogen content are nearly identical to those of the reference a-Si:H films [9]. However, there is a marked difference in hydrogen-bonding configuration between a-Si:H films deposited by the two different techniques. In the FTIR measurement [10], the a-Si:H films grown by triode PECVD exhibit low intensity of the silicon-hydride stretching mode at ~2100 cm^{-1}, which is a signature of the material having a compact microstructure [16]. The Si-H_2 bond density, deduced from the high stretching mode in the FTIR spectrum, is reduced from 2.1 to 0.7 at.%, while the Si-H bond density, deduced from the low stretching mode (~2000 cm^{-1}), remains nearly constant at 10-11 at.%.

Figure 3 shows the degradation characteristics of the 250-nm-thick a-Si:H p-i-n solar cells under 1 sun illumination condition. The absorber layers were grown either by triode or diode PECVD under the conditions that were used for layer depositions described above. Although the initial efficiencies are comparable, long-term light soaking results in markedly different stabilized efficiencies (η_{stb}) between solar cells prepared by triode and diode PECVD. After 1000 h of light exposure, the solar cells prepared by triode PECVD show relative degradation of ~10% in conversion efficiency (defined as degradation ratio $\Delta\eta/\eta_{ini} = (\eta_{ini}-\eta_{stb})/\eta_{ini}$), which is about 10% smaller than the degradation ratio of our reference cell prepared by diode PECVD. As shown in Fig. 3, the high stabilized efficiencies obtained for the solar cells prepared by triode PECVD mainly originate from the low degradation in J_{sc} and FF.

Figure 4 shows the sub gap absorption spectra of a-Si:H measured by FTPS in the actual p-i-n device structure. The FTPS spectra were measured in the initial and light-soaked states for the a-Si:H p-i-n solar cells prepared by the two deposition techniques. The bumps appearing in the sub-band-gap energy region are attributed to optical interference effect. It is clearly shown

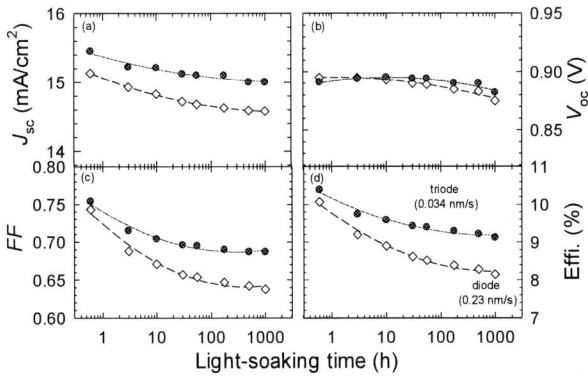

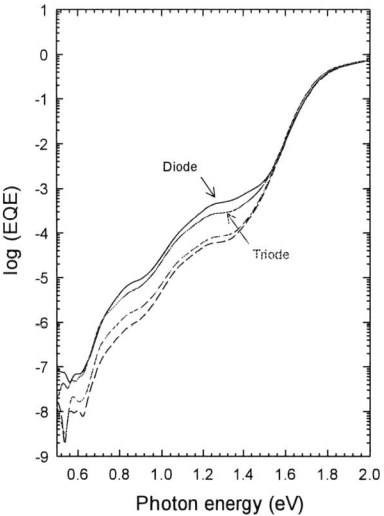

Fig. 3. Stability of the a-Si:H solar cells (i-layer thickness: 250 nm) under long-term light soaking (AM1.5, 100 mW/cm^2, 50 °C, open circuit). The a-Si:H absorbers were prepared by triode (circles) and standard diode (triangles) PECVD techniques.

Fig. 4. FTPS spectra in the initial (dashed lines) and stabilized (solid lines) states for a-Si:H p-i-n solar cells whose absorbers were grown by triode and diode PECVD. The absolute values were corrected using the external quantum efficiency curves of the solar cells.

that the absorption at photon energy of <1.5 eV after light soaking is entirely lower for the a-Si:H solar cell grown by triode than that by diode PECVD, which is consistent with the J-V results of the corresponding solar cells. Interestingly, in contrast, the higher sub gap absorption is found in the initial state, which is likely to have relation to the lower band gap (lower hydrogen content) of the absorber layer grown by triode PECVD. In fact, we observed that the sub gap absorption in the initial state is systematically reduced with increasing the band gap of a-Si:H, e.g., by increasing the hydrogen dilution during the triode PECVD process.

High efficiency a-Si:H single-junction solar cells

Until now, we have used SnO$_2$-coated glass (Asahi-VU) as standard substrate for the fabrication of a-Si:H based solar cells. In addition to the standard VU substrates, we have applied high-haze VU substrates, which were also provided by Asahi Glass Company, to improve the light trapping in the long wavelengths. As shown in Fig. 5, the high-haze VU substrate exhibits the higher diffuse transmission over the entire range of wavelength as compared to the standard VU due to the larger surface roughness. As a result, the external quantum efficiencies (EQE) of solar cells in the long wavelengths is effectively enhanced, resulting in a gain of J_{sc} of ~0.3 mA/cm^2 while keeping V_{oc} and FF almost unchanged. We believe that there is still room for further optimization of substrate texture for better light trapping.

The SnO$_2$ surface was covered with TiO$_2$ (35 nm)-ZnO:Al (10 nm) to improve the light in-coupling at the TCO/Si interface [14]. As shown in Fig. 6, this TiO$_2$-ZnO bilayer coating increases the EQE by 3% in the wavelengths between 400 and 700 nm. Consequently, J_{sc} increases by 0.6 mA/cm^2. Furthermore, the reflection loss is reduced to almost 0% because these samples have an additional AR multilayer coating on glass surface. Nevertheless, the maximum EQE peak is still as low as ~92%. It can be considered that the rest of the incident light (~8%) is either absorbed by the layers other than the i-layer and/or scattered towards the lateral direction through the glass and TCO.

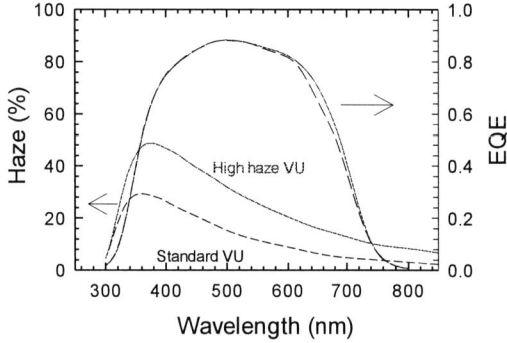 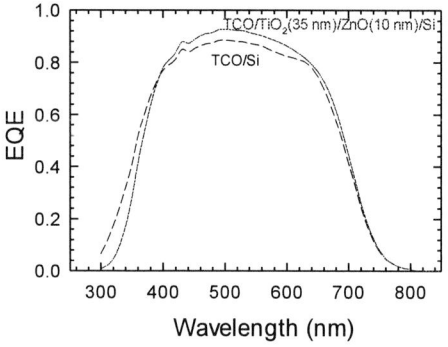

Fig. 5. Haze spectra (diffuse transmittance) of the high-haze and standard Asahi-VU substrates. The EQE spectra of the 250 nm-thick solar cells using these different substrates are also shown.

Fig. 6. EQE spectra of the 250 nm-thick a-Si:H p-i-n solar cells with and without TiO$_2$ (35 nm)-ZnO (10 nm) stacked layers between SnO$_2$ (high-haze VU) and Si interface. The solar cells have an additional antireflection coating on glass surface of the illumination side.

The a-SiC:H p- and a-Si:H buffer layers were designed as thin as possible so that the influence of these layers on the light-soaking stability of solar cell can be minimized. Figure 7 shows the V_{oc} of the a-Si:H solar cells in the initial and stabilized states as a function of the a-SiC:H p-layer thickness. The initial V_{oc} strongly depends on the p-layer thickness because a certain thickness of doped layer is required to build up an internal electric field across the p-i-n junction. Nevertheless, we observed a massive light-induced increase in V_{oc} particularly for the solar cells that have thin p-layers ($t_p < 4$ nm). This effect has already been explained by the increase of the charged light-induced defects near the p-i interface that push the Fermi level

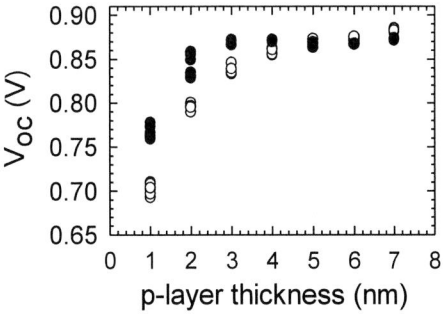

Fig. 7. V_{oc} of a-Si:H p-i-n junction solar cells in the initial (open) and stabilized (filled) states as a function of the a-SiC:H p-layer.

downward to the valence band edge [21]. We found that this effect is even more pronounced when the high-haze TCO substrate is used. Meanwhile, it is widely known that the high band gap buffer layer at p-i interface plays an important role in boosting V_{oc} of a-Si:H solar cells [22-24]. However, a strong degradation is observed when the thickness of a-SiC:H buffer layer is increased [23,24]. To minimize the degradation of solar cell efficiency while keeping V_{oc} relatively high, we deposited a-Si:H as a buffer layer deposited under high H_2 dilution condition.

Figure 8 shows the stability under 1 sun illumination for the a-Si:H solar cells fabricated using an optimum device design described above. The variation of initial J_{sc} and efficiency is caused by the difference of absorber thickness (t_i~190-230 nm). Despite of the performance variation in the initial state, the solar cell parameters become more or less the same after several hundred hours of light soaking. It is seen that almost all solar cell parameters are stabilized after

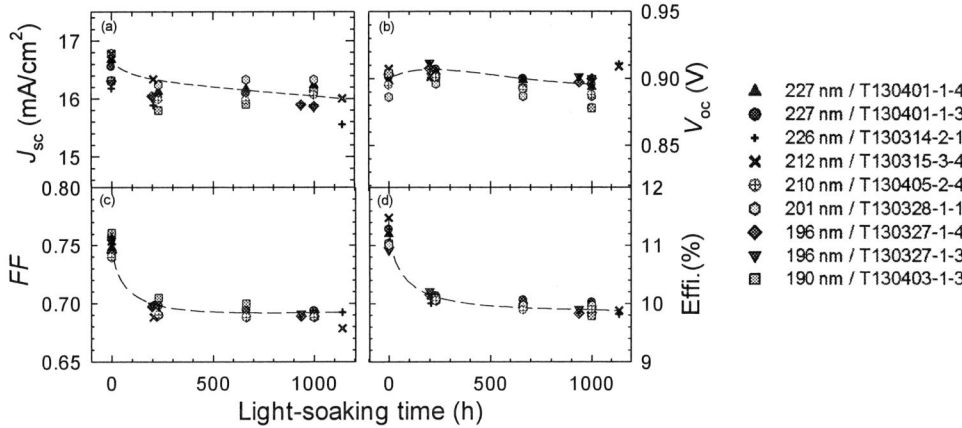

Fig. 8. Degradation characteristics of the a-Si:H solar cells under long-term light soaking (AM1.5, 100 mW/cm², 50 °C, open circuit). The absorber layer thickness indicated in the legend was measured by ellipsometry for p-i-n layer stack on glass substrate prepared in the same deposition run.

Table 1. J-V parameters of an a-Si:H single junction solar cell (i-layer thickness: 227 nm) in the initial and stabilized states (light-soaking condition: AM1.5, 100 mW/cm^2, 50 °C, 1000 hours, open circuit). All measurements were done under designated area illumination (12.5 mm × 8.0 mm) using a black mask.

Sample ID	state	measurement	J_{sc} (mA/cm^2)	V_{oc} (V)	FF	η (%)	area (cm^2)
T130401-1-3	initial	in-house	16.55	0.901	0.757	11.28	1.0
T130401-1-3	stabilized	in-house	16.12	0.896	0.694	10.02	1.0
T130401-1-3	**stabilized**	**AIST**	**16.05**	**0.906**	**0.695**	**10.11**	**0.999**

600 hours of light exposure. Among the several cells, the best performing solar cell (T130401-1-3) shows a stabilized efficiency of as high as 10% after light-soaking for 1000 h. This solar cell was also measured by CSMT-AIST for high accuracy measurement. The measurement results of this solar cell are summarized in Table 1. The results of solar cell performance are in good agreement between two measurements. The relatively large difference is found in V_{oc}, which might be due to the poor control of cell temperature during the in-house measurement. From the high accuracy measurement performed by CSMT-AIST, a stabilized efficiency of 10.11% was confirmed [7]. Note that the cell performance was characterized under designated area AM1.5 illumination using a black mask (1 cm^2) which has a slightly smaller aperture area than the device area (1.04 cm^2). The decrease of J_{sc} with respect to that without mask is in the range of 0.3-0.4 mA/cm^2. This solar cell efficiency is as high as the record stabilized cell efficiency of 10.09% (J_{sc}=17.28 mA/cm^2, V_{oc}=0.876 V, FF=0.665 (confirmed by NREL)) reported by Benagli et al. of TEL Solar (the former Oerlikon Solar) in 2009 [25]. Although the stabilized efficiency is comparable, some notable differences are found by comparing these solar cells. First, J_{sc} of our solar cell is lower by more than 1 mA/cm^2. From the EQE measurement, the low J_{sc} is found to come from the relatively low response in the long wavelength region (λ>600 nm). This difference can be attributed either to a higher band gap of the i-layer or a different light trapping scheme. Second, our solar cell shows higher FF, which can mainly be attributed to the improved light-soaking stability of the a-Si:H absorber layer.

Figure 9 shows the illuminated J-V parameters in the initial and stabilized states of the a-Si:H p-i-n solar cells as a function of the i-layer thickness in the range from t_i=150 to 390 nm. Solar cells were prepared with a full AR design (i.e. AR on glass surface and at TCO/p interface) to evaluate the cell degradation with the high initial performance. In the initial state, J_{sc} increases monotonically with increasing thickness, while V_{oc} and FF tend to decrease. Consequently, the initial efficiency increases gradually with increasing i-layer thickness. After the light soaking, solar cell parameters are degraded except for the V_{oc}. In particular,

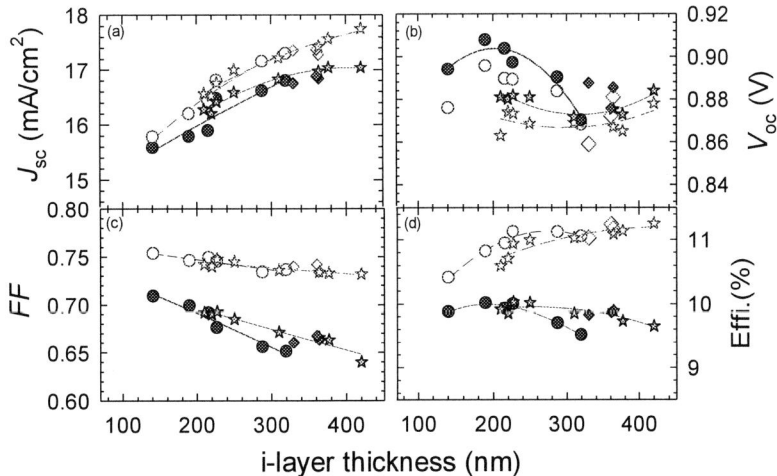

Fig. 9. J-V parameters of the a-Si:H solar cells in the initial (open symbols) and stabilized (filled symbols) states as a function of the i-layer thickness. Results of the solar cells prepared with a modified electrode design in the triode PECVD are also shown (stars). The further improvement was made by optimizing the deposition condition (diamonds).

degradation of FF becomes larger as the i-layer thickness increases. As a result, we found a broad maximum of stabilized efficiency at t_i~220 nm. Recently, we found that the higher efficiencies can be maintained for thicker cells when a modified electrode design is applied in the triode PECVD. It is worth mentioning that these solar cells maintain high stabilized efficiencies (η_{stb}~9.8%) even when the thickness of the i-layer is increased to t_i~370 nm. Moreover, it has been confirmed by CSMT-AIST that a cell with t_i~310 nm has a stabilized efficiency of 10.08%. The high stabilized efficiency a-Si:H solar cell at a greater cell thickness is of particular importance for its application to tandem solar cells because of the requirement of the high top-cell current in improving the tandem solar cell efficiency. Further study is necessary to understand how the plasma condition of the new electrode design and deposition rate influence the light-soaking stability of a-Si:H solar cells.

CONCLUSIONS

A triode electrode configuration is applied in the PECVD process to grow high quality a-Si:H i-layer for the fabrication of p-i-n junction solar cells. The a-Si:H single-junction solar cells exhibit low light-induced degradation of conversion efficiency ($\Delta\eta/\eta_{ini}$~10-12% at t_i=200-250 nm), and their stability is less sensitive to the i-layer thickness in comparison with the high-efficiency solar cells previously reported. By performing the device optimization including

TCO, AR layers and doped layers, we have attained a confirmed stabilized efficiency of 10.11% and 10.08% for 220 nm and 310 nm-thick a-Si:H single-junction solar cells, respectively.

ACKNOWLEDGMENTS

The authors are thankful to Mr. Miyagi, Murata, Sato and Ms. Hozuki for technical assistance, and to Dr. Hishikawa, Ms. Sasaki and Moriya in CSMT-AIST for high accuracy measurements, and to technical board member of PVTEC for useful discussion. They also thank researchers in EPFL and Delft University of Technology for technical discussion on FTPS measurement. This work was supported by the New Energy and Industrial Technology Development Organization (NEDO), Japan.

REFERENCES

1. K. Yamamoto, A. Nakajima, M. Yoshimi, T. Sawada, S. Fukuda, T. Suezaki, M. Ichikawa, Y. Koi, M. Goto, T. Meguro, T. Matsuda, M. Kondo, T. Sasaki, and Y. Tawada, Sol. Energy 77, 939 (2004).
2. M. Boccard, C. Battaglia, S. Hänni, K. Söderström, J. Escarré, S. Nicolay, F. Meillaud, M. Despeisse, and C. Ballif, Nano Lett. 12, 1334 (2012).
3. U. Kroll, J. Meier, L. Fesquet, J. Steinhauser, S. Benagli, J.-B. Orhan, B. Wolf, D. Borrello, L. Castens, Y. Djeridane, X. Multone, G. Choong, D. Domine, J.-F. Boucher, P.-A. Madliger, M. Marmelo, G. Monteduro, B. Dehbozorgi, D. Romang, E. Omnes, M. Chevalley, G. Charitat, A. Pomey, E. Vallat-Sauvain, S. Marjanovic, G. Kohnke, K. Koch, J. Liu, R. Modavis, D. Thelen, S. Vallon, A. Zakharian, and D. Weidman, Proc. 26th European Photovoltaic Solar Energy Conf./Exhib., 2011, p. 2340.
4. A. Terakawa, M. Hishida, S. Yata, W. Shinohara, A. Kitahara, H. Yoneda, Y. Aya, I. Yoshida, M. Iseki, and M. Tanaka, Proc. 26th European Photovoltaic Solar Energy Conf./Exhib., 2011, p. 2362.
5. N. Kadota, M. Hino, A. Tanimoto, K. Murakami, M. Fukuda, W. Yoshida, T. Sasaki, S. Fukuda, T. Nomura, and A. Nakajima, Tech. Dig. 21st Int. Photovoltaic Science and Engineering Conf., 2011, 2A-2O-05.
6. B. Stannowski, O. Gabriel, S. Calnan, T. Frijnts, A. Heidelberg, S. Neubert, S. Kirner, S. Ring, M. Zelt, B. Rau, J.-H. Zollondz, H. Bloess, R. Schlatmann, B. Rech, Sol. Energy Mater. Sol. Cells 119, 196 (2013).
7. T. Matsui, H. Sai, T. Suezaki, M. Matsumoto, K. Saito, I. Yoshida, and M. Kondo, Proc. 28th European Photovoltaic Solar Energy Conference and Exhibition, 2013, p. 2213.
8. D. L. Staebler and C. R. Wronski, Appl. Phys. Lett. 31, 292 (1977).

9. T. Matsui, H. Sai, K. Saito, M. Kondo, Jpn. J. Appl. Phys. 51, 10NB04 (2012).

10. T. Matsui, H. Sai, K. Saito, M. Kondo, Prog. Photovolt: Res. Appl. 21, 1363 (2013).

11. M. Vaneček and A. Poruba, Appl. Phys. Lett. 80, 719 (2002).

12. J. Holovskú, A. Poruba, Z. Purkrt, M. Vaneček, J. Non-Cryst. Solids, 354, 2167 (2008).

13. J. Melskens, G. van Elzakker, Y. Li, and M. Zeman, Thin Solid Films, 516, 6877 (2008).

14. T. Fujibayashi, T. Matsui, and M. Kondo, Appl. Phys. Lett. 88, 183508 (2006).

15. W. Luft, B. von Roedern, B. Stafford, and L. Mrig, Proc. 23rd IEEE Photovoltaic Specialists Conf., 1993, p. 860.

16. E. Bhattacharya and A. H. Mahan, Appl. Phys. Lett. 52, 1587 (1988).

17. T. Nishimoto, M. Takai, H. Miyahara, M. Kondo, and A. Matsuda, J. Non-Cryst. Solids 299-302, 1116 (2002).

18. A. Matsuda, T. Kaga, H. Tanaka, and K. Tanaka, J. Non-Cryst. Solids 59-60, 687 (1983).

19. S. Shimizu, M. Kondo, and A. Matsuda, J. Appl. Phys. 97, 033522 (2005).

20. H. Sonobe, A. Sato, S. Shimizu, T. Matsui, M. Kondo, and A. Matsuda, Thin Solid Films 502, 306 (2006).

21. P. Siamchai, and M. Konagai, Proc. 25th IEEE Photovoltaic Specialists Conf., 1996, p. 1093.

22. K.S. Lim, M. Konagai, and K. Takahashi, J. Appl. Phys. 56, 538 (1984).

23. H. Sakai, T. Yoshida, S. Fujikake, T. Hama, and Y. Ichikawa, J. Appl. Phys. 67, 3494 (1990).

24. B. Rech, C. Beneking, and H. Wagner, Sol. Energy Mater. Sol. Cells 41-42, 475 (1996).

25. S. Benagli, D. Borrello, E. Vallat-Sauvain, J. Meier, U. Kroll, J. Hoetzel, J. Bailat, J. Steinhauser, M. Marmelo, G. Monteduro, and L. Castens, Proc. 24th European Photovoltaic Solar Energy Conf., 2009, p. 21.

Mater. Res. Soc. Symp. Proc. Vol. 1666 © 2014 Materials Research Society
DOI: 10.1557/opl.2014.921

Minority Carrier Annihilation at Crystalline Silicon Interface in Metal Oxide Semiconductor Structure

Jun Furukawa[1], Satoshi Shigeno[1], Shinya Yoshidomi[1], Tomohito Node[1], Masahiko Hasumi[1], Toshiyuki Sameshima[1], and Tomohisa Mizuno[2]

[1]Tokyo University of Agriculture and Technology, Tokyo, 184-8588 Japan
[2]Kanagawa University, Kanagawa, 259-1293 Japan

ABSTRACT

We report photo induced minority carrier annihilation at the silicon surface in a metal–oxide–semiconductor (MOS) structure using 9.35 GHz microwave transmittance measurement. 7 Ωcm n-type 500-μm-thick crystalline silicon substrate coated with 100-nm-thick thermally grown SiO_2 layers was used. 0.2-cm-long Al electrode bars were formed at the top and rear surfaces. 635 nm light illumination onto the top surface caused photo induced carriers to be in one side of the silicon region of the Al electrode. Microwave transmittance system detected photo induced carriers diffused from the light illuminated region via the MOS structured region. When the bias voltage was applied at +2.0 and -2.2 V to the electrode at the top surface, the surface recombination velocity increased from 44 (initial) to 83 and 86 cm/s, respectively because of depletion region formation at rear and top surface respectively. Those voltage applications caused change in the distribution of photo induced carriers in a 0.6-cm-wide region including light illuminated, MOS structured, microwave irradiated regions.

INTRODUCTION

A long carrier lifetime with a low density of defect states is required for fabrication of high performance devices such as photo sensors, solar cells and metal-oxide-semiconductor (MOS) field effect transistors [1-3]. Carrier recombination defect states seriously affect the minority carrier effective lifetime of silicon [4,5] and mainly concentrate at the silicon surface or its interface. In the case of MOS structure, the application of a bias voltage onto metal electrodes causes band bending and changes in the carrier concentration of the silicon surface region that satisfy the charge neutrality between both sides of the oxide insulating layer. Because the bias voltage changes the occupation probability of surface defect states, the photo induced carrier recombination probability depends on the bias voltage. The direct measurement of photo induced carrier recombination probability as a function of bias voltage will give useful information for fabricating high performance photo sensors and passivation layers for photovoltaic devices. Recently, we reported the measurement of the light induced excess carrier recombination velocity at the silicon surface in the MOS structure using a 9.35 GHz microwave transmittance measurement system with continuous wave (CW) light illumination [6,7]. We demonstrated a change in the microwave absorption of photo induced carriers laterally diffused in the silicon region of the MOS structure with the application of bias voltage. The bias voltage application causes a change in the carrier recombination probability.

In this paper, we report precise analysis of distribution of photo-induced carrier density in the lateral direction from light illuminated, MOS structured, microwave irradiated regions. We show voltage application change in carrier recombination velocity at the SiO_2/Si surface and change distribution of photo-induced carrier density over a wide range of 0.6 cm. We also discuss the optimum condition of our method for measuring field induced annihilation of photo induced carriers.

EXPERIMENT

7 Ωcm n-type crystalline silicon substrate with a thickness of 500 μm was prepared. Its top and rear surfaces were coated with 100-nm-thick thermally grown SiO_2 layers. In advance of the present investigation, we confirmed the majority electron carrier density and effective photo-induced minority carrier lifetime τ_{eff} of 6.0 x 10^{14} cm^{-3} and 1.1 x 10^{-3} s, respectively using our conventional equipment [6]. A set of 100-nm-thick Al electrodes with a size of 0.2 x 4.0 cm^2 were formed on the top and rear surfaces each by vacuum evaporation. Al electrodes faced each other in the same horizontal position. A 9.35 GHz microwave transmittance measurement system was used in order to detect the photo induced minority carriers of the samples, as shown in Fig. 1(a). The system had waveguide tubes, which had a narrow gap. The sample was placed in the gap between the waveguide tubes. CW 635 nm light was introduced on to the outside of waveguide tube with a cross section of 1.0 x 2.3 cm^2 when top and rear Al electrodes were coincidentally positioned on the 0.2-cm-thick wall of the waveguide tubes, as shown in Fig. 1(b). The light intensity was set at 1.5 mW/cm^2 at the sample surface. The microwave transmissivities of the sample in the dark, T_d, and under light illumination, T_p, were measured for duration of 4 s each [6].

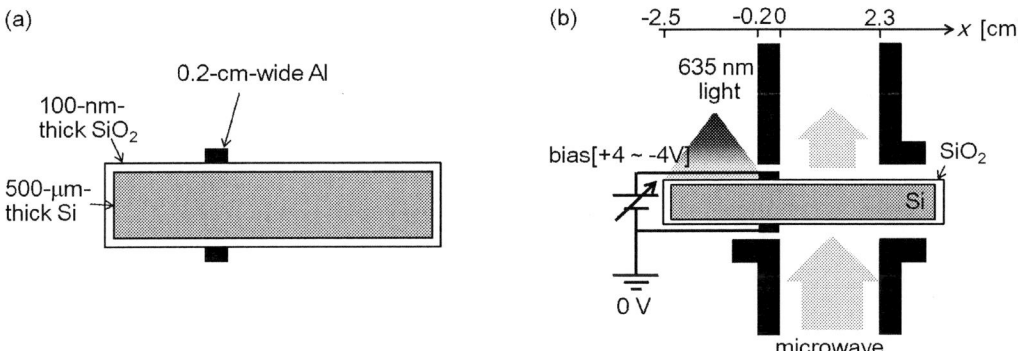

Fig. 1. The schematic structure of Al/SiO$_2$/Si/SiO$_2$/Al sample (a) and 9.35 GHz microwave transmittance measurement system with bias voltage application to top electrode while keeping rear electrode at 0 V (b).

Photo induced carriers with long τ_{eff} traverse under the 0.2-cm-thick wall of the waveguide tube and diffuse into the region of microwave absorption measurement. Photo induced carriers that laterally diffused into the region in the waveguide tube via the region covered with the Al electrodes were observed with microwave absorption. $\ln(T_d/T_p)$ was measured by turning the CW 635 nm light ON and OFF. Moreover, the microwave absorption measurement was carried out by applying a bias voltage from -4 to 4 V to top electrode and keeping rear electrode at 0 V. Each bias voltage was maintained for 10 s, in which T_d was measured for 4 s at 2 s from the bias voltage application and T_p was subsequently measured for 4 s in order to realize measurement in the electrically steady state.

In order to analyze the change in microwave transmissivities, a numerical analysis program of carrier diffusion and annihilation using the finite differential element method was constructed. In the simple carrier diffusion model, the surface recombination velocity was programmed to be variable for the region of MOS structure. The photo induced minority carrier density $N(x, y)$ as a function of the lateral position x and the depth position y from the top surface is given using the carrier diffusion model under a two-dimensional steady-state condition as [8]

$$D\frac{\partial^2 N(x,y)}{\partial x^2} + D\frac{\partial^2 N(x,y)}{\partial y^2} - \frac{N(x,y)}{\tau_b} + g(x,y) = 0 \tag{1a}$$

$$D\frac{\partial N(x,y)}{\partial y}\bigg|_{y=0} = S_{top}N(x,0) - g(x,0)\Delta y \tag{1b}$$

$$D\frac{\partial N(x,y)}{\partial y}\bigg|_{y=d} = S_{rear}N(x,d) - g(x,d)\Delta y \tag{1c}$$

,where D is the diffusion coefficient of the minority carrier, τ_b is the bulk lifetime assumed to be large at 0.01 s, d is the substrate thickness of 500 μm, $g(x, y)$ is the carrier generation rate, and Δy is the lattice unit in the depth direction of 20 μm. The lattice unit in the horizontal direction, Δx, was also set at 20 μm. Calculation boundary conditions were placed corresponding to the experimental configuration, as shown in Fig. 2. $g(x, y)$ was only set as a constant of g at the x position lower than -0.2 cm and at the first lattice of the depth for carrier generation at the surface outside of the waveguide tube. The total minority carrier number N_T in the region where the microwave transmitted the sample was then calculated by integrating $N(x, y)$ with respect to y from 0 to d and to x from 0 to 2.3 cm (the length of the microwave tube). $\ln(T_d/T_p)$ was calculated from N_T using the free-carrier microwave absorption theory [7]. The most possible S_T ($= S_{top} + S_{rear}$) in the region with x between -0.2 and 0 cm was determined by fitting the calculated $\ln(T_d/T_p)$ values to the experimental ones under assumptions of S_{rear} constant of initial value of 22 cm/s for negative voltage application, and S_{top} constant of initial value of 22 cm/s for positive voltage application. We also calculated distribution of photo-induced carrier density in the lateral direction from light illuminated, MOS structured, microwave irradiated regions with different bias voltages.

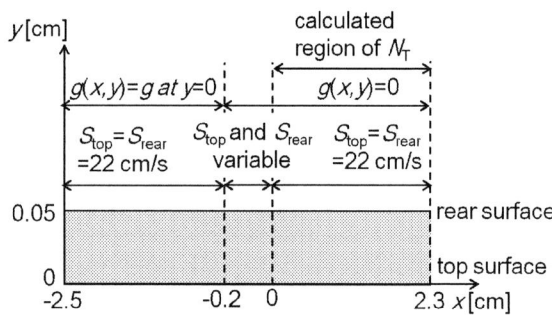

Fig. 2. Two dimensional numerical calculation model of photo induced carrier diffusion in the lateral x and depth y directions.

RESULTS AND DISCUSSION

From the initial τ_{eff} of 1.1 x 10^{-3} s, S_T was obtained as 44 cm/s determined by our conventional analysis [6]. The silicon surfaces were well passivated by thermally grown SiO$_2$ under the bias-free condition. The effective minority hole carrier diffusion length L was estimated as 0.115 cm. This large L indicates that the photo induced excess minority carriers diffused over the entire substrate thickness and over a long horizontal distance. $\ln(T_d/T_p)$ was 3.4 x 10^{-3} at 0 V in the bias case of top and rear electrodes, which was the same as that in the open-circuit case. We interpret the generation of photo induced minority carriers by 635 nm light illumination diffused in the lateral direction into the area of the waveguide tube with an initial τ_{eff} of 1.1 x 10^{-3} s at a bias voltage of 0 V. $\ln(T_d/T_p)$ decreased from 3.4 x 10^{-3} to 1.9 x 10^{-3} as the bias voltage increased from 0 to +2.0 V and decreased to 1.8 x 10^{-3} as the bias voltage decreased to -2.2 V. This indicates that carrier annihilation rate in the region covered with the Al electrode was increased by the bias voltage application to +2 V. On the other hand, $\ln(T_d/T_p)$ increased again to 2.7 x 10^{-3} and 2.5 x 10^{-3} as the bias voltage further increased to +4 V and decreased to -4 V, respectively. This means that the carrier annihilation rate in the region underlying the Al electrode became low again at a high bias voltage.

Figure 3 shows the S_T analyzed from the experimental $\ln(T_d/T_p)$ as a function of the bias voltage. S_T symmetrically changed with the bias voltage in the case of bias voltage application to the top electrode. It increased from 44 to 83 cm/s as the bias voltage increased from 0 to +2.0 V. It also increased to 86 cm/s as the bias voltage decreased to -2.2 V. On the other hand, S_T decreased to 57 cm/s as the bias voltage further increased to +4 V. It decreased again to 63 cm/s as the bias voltage further decreased to -4 V.

The numerical analysis revealed that the decrease in N_T was caused by increase in S_T, as shown in Fig. 3. When a positive bias voltage is applied to top electrode for the sample structure Al/SiO$_2$/Si/SiO$_2$/Al, the surface potential at the silicon top surface below top electrode decreases and an electron-accumulating region is formed. In contrast, the surface potential at the silicon rear surface above rear electrode moves up and electron depleted region is formed at the rear surface. On the other hand, the negative bias voltage application causes the opposite situation with electron depleting at the top surface and electron accumulating at the rear surface. When light-induced excess minority carriers traverse at the carrier depletion edge near the silicon surface, minority carriers will be accelerated toward the silicon surface owing to the attractive potential slope. Carrier recombination will be promoted by minority carrier injection into the surface associated with the potential slope in the carrier depletion region [8].

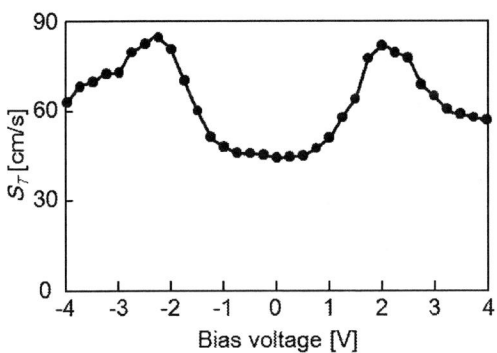

Fig.3. S_T analyzed from experimental $\ln(T_d/T_p)$ as a function of the voltage biased for top electrode while keeping rear electrode at 0 V.

Figure 4 shows the calculated carrier density $N(x)$ obtained by integration of $N(x, y)$ with the depth y from the top to rear surfaces as a function of the lateral position x including light illuminated, MOS structured, and microwave irradiated regions. $N(x)$ monotonously decreased as x increased even in the light illuminated region because of diffusion of photo-induced carrier in the lateral direction. While it further

decreased in MOS structured region, there was residual $N(x)$ in the microwave irradiated region. When S_{top} and S_{rear} were assumed to be 64 and 22 cm/s appeared in the case of -2.2 V bias voltage application at the MOS structured region, $N(x)$ had lower values in the 0.6 cm-wide region shown in Fig. 4 than those in the case of S_{top} and S_{rear} of 22 cm/s (zero voltage application case). Carrier annihilation increased in the wide region owing to high carrier diffusion.

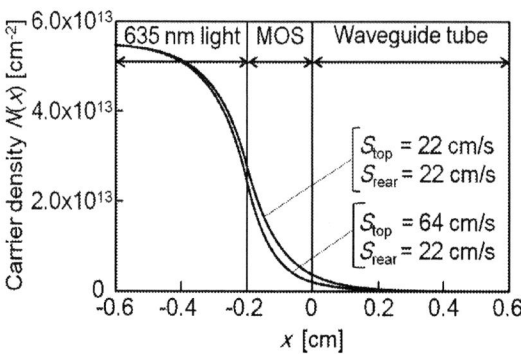

Fig. 4. Calculated carrier density $N(x)$ obtained by integration with the depth y from the top to rear surfaces as a function of the lateral position x including light illuminated, MOS structured, and microwave irradiated regions increasing wise of x.

Figure 5 shows the calculated N_T difference N_{diff} as a function of the metal length L, in which the region of MOS structure ranges from $-L$ and 0 cm in the x direction. N_{diff} were calculated by N_T with S_{top} and S_{rear} of 22 cm/s subtracted by N_T with S_{top} and S_{rear} of 64 and 22 cm/s. N_{diff} increased and had peaks at a metal length of 0.08 cm. it monotonously decreased as the metal length increased above 0.08 cm. The results of N_{diff} indicate that 0.08-cm metal length was the most sensitive for observing field effect photo induced carrier annihilation when the S_T changes by 42 cm/s from 44 to 86 cm/s. Our present experimental metal length of 0.2 cm was larger than the best length of 0.08 cm, but it was small enough to detect field effect photo induced carrier annihilation because our

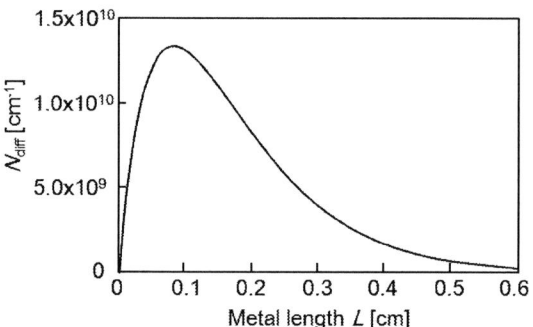

Fig. 5. Calculated N_{diff} as a function of the metal length of the MOS structure.

experimental accuracy of N_{diff} of 1.0×10^9 cm^{-1} allows L ranging from 0.01 to 0.45 cm in this condition. The N_{diff} result of Fig.5 is given in the case of the high τ_{eff} of 1.1×10^{-3} s of the present sample, which gives a long diffusion length of the minority carrier of 0.117 cm and the most sensitive L of 0.08 cm. Defective or higher doped sample can have low τ_{eff}. For example, if τ_{eff} is 1.0×10^{-4} s, typical value of solar grade silicon, the diffusion length of the minority carrier decreases to 0.035 cm and the most sensitive L is 0.032 cm if the surface recombination velocity was changed by 44 cm/s. Our experimental accuracy of N_{diff} of 1.0×10^9 cm^{-1} gives measurement limit of τ_{eff} of 6.3×10^{-5} s in the present condition. These results indicate that the present experimental system can give information on carrier annihilation property in MOS region with bias voltage application.

CONCLUSIONS

We reported photo induced minority carrier annihilation properties at the silicon surface in a MOS structure using a 9.35 GHz microwave transmittance measurement system. 7 Ωcm n-type 500-μm-thick crystalline silicon substrates coated with 100-nm-thick thermally grown SiO_2 layers were prepared. The initial samples had τ_{eff} and S_T of 1.1 x 10^{-3} s and 44 cm/s, respectively. A set of Al electrodes with a length of 0.2 cm were formed on top and bottom surfaces each. We observed $\ln(T_d/T_p)$ as a function of the bias voltage with the structure Al/SiO_2/Si/SiO_2/Al under light illumination outside of the waveguide tubes. S_T increased from 44 to 83 cm/s as the bias voltage increased from 0 to +2.0 V and also increased to 86 cm/s as the bias voltage decreased to -2.2 V because the depletion regions were formed at rear and top surfaces, respectively. Those voltage application caused change in the distribution of photo induced carriers top surface a 0.6-cm-wide region including light illuminated, MOS structured, microwave irradiated regions. The optimum length of the electrode was estimated as 0.08 cm as the most sensitive conditions. Our measurement accuracy allows the metal length from 0.01 to 0.45 cm.

ACKNOWLEDGEMENTS

This work was partly supported by a Grant-in-Aid for Science Research C (Nos. 25420282 and 23560360) from the Ministry of Education, Culture, Sports, Science and Technology of Japan, and Sameken Co., Ltd.

REFERENCES

1. S. M. Sze, Semiconductor Devices (Wiley, New York, 1985) Chap. 7.
2. C. D. Arvanitis, S. E. Bohndiek, G. Royle, A. Blue, H. X. Liang, A. Clark, M. Prydderch, R. Turchetta, and R. Speller, Med. Phys. 34, 4612 (2007).
3. M. A. Green, K. Emery, Y. Hishikawa, W. Warta, and E. D. Dunlop, Prog. Photovoltaics 20, 12 (2012).
4. G. S. Kousik, Z. G. Ling, and P. K. Ajmera,J. Appl. Phys. 72, 141 (1992).
5. K. Sakamoto and T. Sameshima, Jpn. J. Appl. Phys. 39, 2492 (2000).
6. T. Sameshima, H. Hayasaka, and T. Haba, Jpn. J. Appl. Phys. 48, 021204 (2009).
7. T. Sameshima, J. Furukawa, T. Nakamura, S. Shigeno, T. Node, S. Yoshidomi, and M. Hasumi, Jpn. J.Appl. Phys. 53, 031301 (2014).
8. W. Shockley and W. T. Read, Jr., Phys. Rev. 87, 835 (1952).

Mater. Res. Soc. Symp. Proc. Vol. 1666 © 2014 Materials Research Society
DOI: 10.1557/opl.2014.666

Direct gap Group IV semiconductors for next generation Si-based IR photonics

John Kouvetakis[1], James Gallagher[2] and José Menéndez[2]

[1] Department of Chemistry and Biochemistry, Arizona State University,
Tempe, AZ 85207, U.S.A.
[2] Department of Physics, Arizona State University,
Tempe, AZ 85207, U.S.A.

ABSTRACT

This paper presents synthesis and optical properties of mono-crystalline $Ge_{1-y}Sn_y$ and $Ge_{1-x-y}Si_xSn_y$ semiconductor alloys grown on Si/Ge platforms via purposely designed CVD routes using highly reactive Si/Ge/Sn hydrides including Ge_3H_8, Ge_4H_{10}, Si_4H_{10} and SnD_4. The $Ge_{1-y}Sn_y$ materials are shown to exhibit strong and tunable photoluminescence induced by the substitution of sizable Sn concentrations in the Ge diamond lattice ultimately leading to an indirect-to-direct band gap crossover at $y = 0.08\text{-}0.09$. The optical data indicate that the IR coverage of the alloy extends well beyond that of elemental Ge into the broader long wavelength range suggesting a variety of applications in Si-based photonics. $Ge_{1-x-y}Si_xSn_y$ alloys represent the first viable ternary semiconductor among group IV elements with independently tunable lattice parameter and electronic structure. Studies of the compositional dependence of direct and indirect edges in these alloys using photoluminescence and photocurrent measurements are reviewed. The optical results show band gap variation over a wide range above and below that of Ge from 1.1 to 0.5 eV and provide the first demonstration of direct gap behavior in this semiconductor system.

INTRODUCTION

The successful epitaxial growth and stabilization of diamond-cubic α-Sn in the early 1980's[1] raised expectations that crystalline $Ge_{1-y}Sn_y$ alloys might also achievable by epitaxial stabilization on suitable substrates.[2,3] This system had attracted considerable attention since the unraveling of the semimetallic band structure of α-Sn,[4] because a simple interpolation between α-Sn and Ge suggested that the alloy, unlike the parent compounds, should be a direct gap semiconductor over a broad compositional range.[5] Subsequent theoretical calculations within the virtual crystal approximation gave results similar to the linear interpolation, predicting a direct band gap for $0.2 < y < 0.6$.[6,7]

Epitaxial stabilization and non-equilibrium growth are critical for $Ge_{1-y}Sn_y$ alloys because the phase diagram of this system reveals a vanishing solid solubility of the two elements.[8] This is in sharp contrast with its fully miscible Ge-Si counterpart, a prototypical alloy system with widespread technological applications.[9] After several attempts with mixed success in the 1987-1997 decade, mostly using molecular beam epitaxy (MBE) and related techniques,[10-13] our group at ASU introduced a CVD approach to $Ge_{1-y}Sn_y$ alloys based on reactions of deuterated stannane (SnD_4), and digermane (Ge_2H_6) at low temperatures between 250°C and 350°C. This method led to films with atomically flat surfaces and high structural quality, as evidenced by atomic force microscopy (AFM), electron microscopy, x-ray, and Rutherford Backscattering (RBS) studies. In particular, the latter show virtually perfect channeling for Ge and Sn and a uniform Sn distribution, demonstrating the complete substitutionality of Sn in the tetrahedral lattice and the elimination of Sn-segregation effects that were a serious concern in prior studies.

Moreover, the CVD route allows the growth of $Ge_{1-y}Sn_y$ alloys directly on Si wafers, which has obvious technological benefits and facilitates the characterization of the film material due to its large contrast with Si. Films with thicknesses in the hundreds of nanometers and approaching 1 µm in some cases became routinely available, and this enabled very detailed studies of the optical, structural, and thermal properties of $Ge_{1-y}Sn_y$ alloys.[14-17] Additional growth refinements made it possible to improve the structural quality and composition accuracy,[18] and to introduce dopants.[19-22] These advances finally enabled the observation of photoluminescence (PL) from $Ge_{1-y}Sn_y$ alloys on Si,[23] as well as the fabrication of heterostructure *pin* diodes with extended infrared responsivity[24-26] and sizable electroluminescence signal.[27]

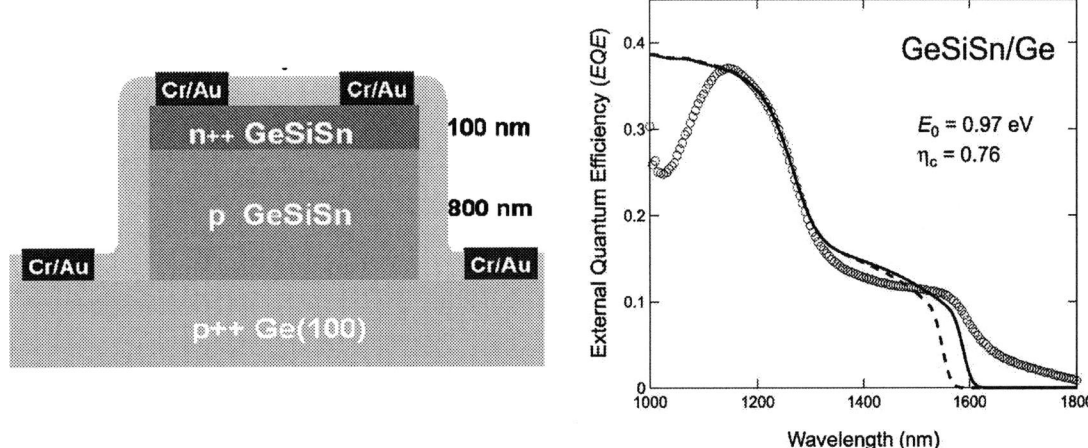

Figure 1: (Left) Schematic of a GeSiSn diode grown on *p*-type Ge wafers. (Right) Circles show the measured external quantum efficiency for the case of a $Ge_{0.87}Si_{0.11}Sn_{0.023}$ active layer. The solid line is a theoretical fit using a model that includes the band gap narrowing in the Ge wafer due to high doping. The dashed line shows the same fit assuming no band gap narrowing in Ge. The measured ternary band gap and collection efficiency for this device are E_0=0.97eV and η_c=0.76, respectively [R. Beeler, D. J. Smith, J. Menendez and J. Kouvetakis, *Photovoltaics IEEE Journal of* **3(4)** 434-440 (2012)].

An extension of the Ge_2H_6 /SnD_4 CVD method led to the introduction of the ternary $Ge_{1-x-y}Si_xSn_y$ system,[28-31] using silylgermane ($SiGeH_6$) or trisilane (Si_3H_8) as Si sources. The CVD growth compatibility between the binary and ternary alloys, combined with the two compositional degrees of freedom in the ternary (which makes it possible to decouple lattice constant and band gap[32]), provided new opportunities for device design and strain/band gap engineering.[33,34] A possible use of the ternary system is the systematic application of tensile strain on Ge or $Ge_{1-y}Sn_y$ layers.[35] Tensile strain reduces the energy separation between the direct and indirect band gaps in Ge and is critical for the operation of interband lasers based on this material. In current laser devices the tensile strain develops when the Ge-on-Si films cool down to room temperature following growth.[36] This method, based on the thermal expansion mismatch between Ge and Si, is difficult to control and cannot be used to generate strains above 0.3%. On the other hand, relaxed $Ge_{1-x-y}Si_xSn_y$ grown on Si substrates can have their Sn and Si components adjusted so that the alloy has a lattice constant greater than that of Ge and a larger band gap, providing at the same time a fully controllable source of tensile strain and a carrier confinement barrier to improve laser action. On the other hand, if the Si/Sn ratio is kept near a value close to

4:1, the system remains lattice-matched to Ge and its optical band gap can be tuned between 0.8 eV and 1.2 eV.[32] Thus a ternary $Ge_{1-x-y}Si_xSn_y$ alloy is an intriguing candidate as the long sought 1-eV-gap material that may be incorporated as a fourth junction to increase the performance of Ge/InGaAs/InGaP multijunction devices.[37] The recent demonstration of $Ge_{1-x-y}Si_xSn_y$ *pn* diodes with excellent electrical and optoelectronic properties (see Figure 1) is a significant breakthrough in this field. A third potential application of $Ge_{1-x-y}Si_xSn_y$/Ge and other lattice-matched multilayer systems is in the area of quantum cascade laser structures.[38] Up to now these devices are based on expensive III-V compounds and substrates.[39] The $Ge_{1-x}Si_x$/Si system, a much more attractive alternative from the point of view of cost, cannot be used for this application due to strain management issues and an accidentally small conduction band offset. By contrast $Ge_{1-x-y}Si_xSn_y$/Ge can be made strain free and they are expected to have a considerable conduction band offset.[31]

The successes of the Ge_2H_6 /SnD_4 route have led to renewed interest in Sn-containing semiconductors. The MBE[8,40-43] and laser-assisted deposition methods have been revisited,[44,45] and in the CVD realm, our collective work has generated significant industrial activity to produce the SnD_4 compound in semiconductor-grade purity for commercial applications. Efforts are under way by several industrial entities to manufacture large quantities of the material and stabilize various concentrations mixed with high purity H_2. The commercial availability of SnD_4 is significant in that it will facilitate the deployment of our growth methods and promote their adoption in both basic research as well as industrial manufacturing of semiconductors.

In spite of the above promising advances, one remaining significant limitation of all methods proposed so far to grow $Ge_{1-y}Sn_y$ is the difficulty in depositing thick films (~ 1 μm) with $y > 0.04$ directly on Si substrates. Recent work confirms[14,46] that the indirect-to-direct crossover in the alloy may take place for Sn concentrations as low as $y = 0.06$, much less than the original predicted threshold of $y = 0.2$, but even these modest concentrations have been extremely difficult to achieve in thick films. This precludes the observation of light emission, which in films with thicknesses below 500 nm is largely suppressed by non-radiative recombination at the intrinsically defected Si/GeSn interface.[47] In the case of the Ge_2H_6 /SnD_4 method, one can achieve thick films (0.5-1 μm) with $y \leq 0.03$ at growth temperatures between 370 °C-340 °C. However, to produce materials with higher Sn concentrations the temperature must be lowered from 340 °C ($y = 0.03$) to 290 °C, ($y = 0.09$) to ensure full substitution of the Sn atoms in the structure. This reduction in temperature results in a concomitant decrease in growth rate, and prevents the film from completely relaxing the mismatch strain with the Si during deposition, particularly for samples with $y > 0.04$.[17] The reduced growth rates and residual compressive strains accumulated during growth limit the overall thickness that can be achieved, ultimately diminishing the device potential of these materials on Si. This limitation reduces the appeal of $Ge_{1-y}Sn_y$ alloys as an alternative to Ge for lasers integrated on Si substrates.[36] Whereas in the case of pure Ge the band bap tuning requires an extrinsic tensile strain provided by the thermal expansion mismatch with the substrate, in $Ge_{1-y}Sn_y$ alloys the Sn concentration provides a more convenient intrinsic tool to control the laser emission wavelengths, but this requires thick films with Sn concentrations at least approaching the indirect to indirect crossover.

In the case of $Ge_{1-x-y}Si_xSn_y$ alloys, there are also several major limitations of the UHV-CVD method that prevent the exploration of the full potential of this system: (1) The single source SiH_3GeH_3 with direct Si-Ge bonds must be used instead of Si_3H_3/Ge_2H_6 mixtures for the production of alloys with $y > 0.02$, because its higher reactivity is more compatible with the low temperatures required for growing single phase mono-crystalline structures of these materials.

Unfortunately, the compound is not yet commercially available in sufficient quantities for routine large-scale fabrication and deployment of the desired alloys in device-quality form. (2) Dopant memory effects (i.e. continued incorporation into subsequently grown intrinsic films) require passivation of the reactor walls with Si prior to each growth experiment. (3) Achieving a reproducible control of layer thicknesses at the monolayer level —necessary for quantum-cascade lasers— is difficult due to the batch nature of the UHV-CVD method.[48]

The main purpose of this paper is to describe recent progress in the development of new approaches to grow binary $Ge_{1-y}Sn_y$ and ternary $Ge_{1-x-y}Si_xSn_y$ alloys that are likely to overcome the above described limitations. Highlights of the research accomplishments are presented below focusing on photoluminescence measurements which demonstrate that these new optimal quality alloys exhibit superior optical performance and in the case of $Ge_{1-y}Sn_y$ provide an accurate determination of the composition at which these materials become direct gap semiconductors opening up new opportunities for advanced device fabrication at long wavelengths beyond that of elemental Ge.

RESULTS and DISCUSSION

The news synthetic approaches to $Ge_{1-y}Sn_y$ and $Ge_{1-x-y}Si_xSn_y$ are based on very recent discoveries at Arizona State University that include the following breakthroughs. (a) the introduction of trigermane (Ge_3H_8) instead of Ge_2H_6 as the source of Ge has led to the growth of thick (> 500 nm) films $Ge_{1-y}Sn_y$ films with Sn concentrations as high as y = 0.09, presumably due to better compatibility with the SnD_4 precursor, which requires very low growth temperature. (b) the use of tetragermane (Ge_4H_{10}) and tetrasilane (Si_4H_{10}) as Ge and Si sources, respectively, has led to the growth of $Ge_{1-x-y}Si_xSn_y$ alloys for the first time in a single-wafer reactor. This breakthrough improves the control of multilayer structures and enables the incorporation of higher amounts of Sn, as needed for studies of potential applications in the areas of lasers and solar cells. (c) the use of Ge buffer layers to reduce the starting lattice mismatch between the Si(100) substrate and the film. This allows formation of largely relaxed epilayers exhibiting reduced dislocation densities at the interface and thus substantially lowers non-radiative recombination velocities relative to analogs grown directly on Si.

Trigermane based $Ge_{1-y}Sn_y$ alloys

To demonstrate the viability of Ge_3H_8 as a chemical source, we have recently developed a practical synthesis route to isolate significant amounts of the compound, enabling a thorough characterization and its use in film growth. [49] The synthesis method employs a pyrolysis reaction of Ge_2H_6 at 250 °C —as described by the equation below—to obtain a 50% yield of the pure Ge_3H_8 product (see Figure 2) in a multi gram (~ 10 g) quantity scale.

$$2 Ge_2H_6 \rightarrow Ge_3H_8 + GeH_4$$

The above reaction also generates several grams of Ge_4H_{10}, which was isolated as a mixture of the linear and branched isomers shown in Figure 2. The Ge_4H_{10} compound exhibited a sufficient vapor pressure of 1.6 Torr at room temperature to allow its use as a viable Ge source for the synthesis of $Ge_{1-x-y}Si_xSn_y$ alloys described is subsequent sections.

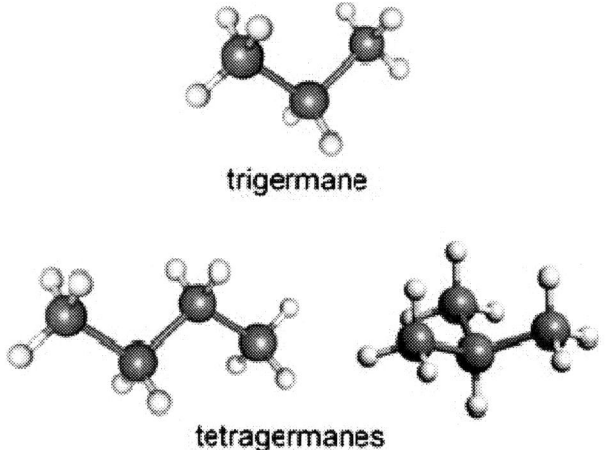

Figure 2: Molecular structures of trigermane (Ge_3H_8) and the two tetragermane (Ge_4H_{10}) isomers produced in this study (see Ref 50).

The compound Ge_4H_{10} was also the essential chemical source for the deposition of Ge buffers on high resistivity 4" Si(100) substrates employed in this study.[50] These buffers typically exhibited large thicknesses (1-2 microns) atomically flat surfaces (AFM RMS <1 nm) and low threading defects densities ($<10^7/cm^2$) making them suitable templates for subsequent growth of $Ge_{1-x-y}Si_xSn_y$ and $Ge_{1-y}Sn_y$ epilayers. The $Ge_{1-y}Sn_y$ epilayers ($y\leq0.11$) were grown on these platforms via UHV CVD reactions of Ge_3H_8 diluted in high purity H_2 and intermixed with SnD_4. The deposition reactor consists of a resistively heated 3"-diameter quartz tube attached to an ultrahigh vacuum chamber. The latter is equipped with a load lock mechanism that allows transferring of the wafer boat into the deposition tube under UHV conditions without compromising the purity of the reaction zone. The substrate boat is fabricated with high-grade fused silica and is configured to accommodate multiple Si wafers with sizes up to 3 inch in diameter. The substrates employed in these studies are quadrants cleaved from the Ge buffered 4" Si(100) wafers. In a typical run the substrates were chemically treated to remove organic contaminants and then dipped in 10% HF/methanol solution to etch the surface oxide, rinsed by methanol, and dried using a nitrogen nozzle. Next, the wafers were loaded onto the boat and inserted into the reactor through the load lock under a flow of the H_2 carrier gas at a pressure of 10^{-4} Torr, while the reactor was kept at the selected growth temperature in the 360-300 °C range. The background pressure of was increased to 0.200 Torr and stabilized over a period of several minutes after which time, the mixture of the SnD_4 and Ge_3H_8 co-reactants was introduced into the H_2 stream and allowed to flow over the substrate surface, initiating the crystal growth process.

The film surface was examined using Nomarski microscopy and found to be optically featureless devoid of defects and imperfections. Atomic Force Microscopy (AFM) images showed a root-mean-square (RMS) roughness of ~ 1-5 nm depending on composition. Cross-Sectional Transmission Electron Microscopy (XTEM) examinations revealed mono-crystalline layers with planar surfaces and diamond-cubic structures. Threading dislocations were observed in the lower segments of the layers above the interface, while the upper portions appeared to be significantly less defective, particularly as the thickness increased and the majority of the film exhibited bulk-like crystal behavior. High-resolution images indicate that the misfit strain between the $Ge_{1-y}Sn_y$ layers and the Si wafer is accommodated by the formation of defects confined to the interface region. These were identified using STEM HAADF images to be Lomer dislocations and short stacking faults penetrating downward into the Ge buffer indicating that the latter provides a low energy platform for the integration of high crystal quality epilayers on to Si wafers.

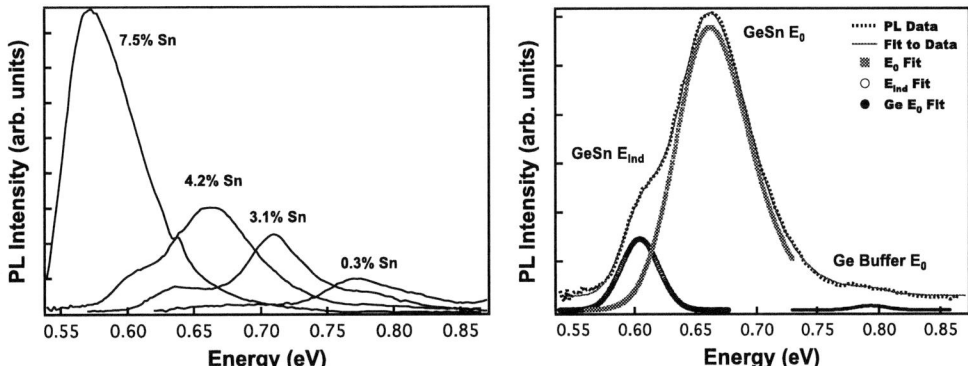

Figure 3: PL spectra of $Ge_{1-y}Sn_y$ samples with y=0.003-0.70. The main peak is assigned to the direct gap emission and the shoulders at the low energy side are attributed to the indirect gaps. Inset shows peak fits of the three main features appearing in the spectrum of a representative y=0.045 sample (direct gap, indirect gap and Ge buffer peak).

Rutherford Backscattering random spectra provided bulk compositions and film thickness estimates in the range of 650-700 nm which are perfectly suitable for PL measurements. The channeled RBS scans were used to ensure that the material was a crystalline, epitaxial and single-phase alloy devoid of Sn precipitates and interstitials. High resolution XRD was used to measure the lattice dimensions indicating that the cubic parameters increase from 5.665 Å to 5.737 Å in the 1-11% Sn composition range. The XRD plots also showed that the "as-grown" layers exhibit residual compressive strains from 0.05% to 0.25% depending on composition. These strains were mostly relaxed by subjecting the samples to rapid thermal annealing treatments.

The availability of high crystal quality samples with Sn concentrations up to $y = 0.11$ indicates that this approach is tantalizingly close to the demonstration of group-IV direct gap materials. In this case, the determination of the indirect to direct gap cross over composition requires photoluminescence measurements, which provide an accurate analysis of both the direct and indirect edge energies as a function of Sn concentration. PL spectra were acquired at room temperature using 400 mW of radiation generated from 980 nm laser focused to a 100 um spot. The emitted light is collected by an f=140 mm Horiba MicroHR spectrometer using a grating with 600 grooves/mm blazed at 2 um. The apparatus is equipped with a liquid nitrogen (LN_2) cooled extended InGaAs detector with a wavelength spectral range of 1300-2300 nm. Under these conditions, the typical spectrum of the $Ge_{1-y}Sn_y$ samples show a strong main peak corresponding to direct gap emission and a lower energy shoulder attributed to indirect transitions. Figure 3 shows representative plots of selected samples with y=0.03-0.070 and similar thicknesses near 650 nm to ensure consistency in the dependence of the PL data on the Sn content. The direct and indirect gap emission features are due to radiative recombination of photo-excited carriers transitioning from the conduction band minima at the Γ and L-points to the Γ point valence band maximum, respectively. From figure 3 it is apparent that as the Sn content is increased the main peak shifts to lower energies and its intensity increases dramatically near the expected cross over composition regime. At the same time, the direct and indirect edges- clearly resolved here for the first time-exhibit a systematic decrease in separation.

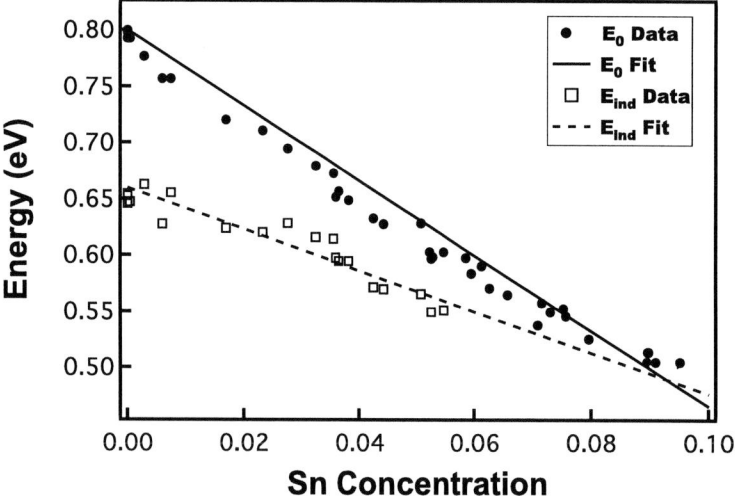

Figure 4: Direct (black dots) and indirect (open squares) band gap energies vs. Sn contents and corresponding linear fits of the data for $Ge_{1-y}Sn_y$ epilayers grown on Ge buffered Si. The direct-indirect crossover is expected to be $y_c \sim 0.09$.

Note that for the y=0.07 sample, a shoulder corresponding to the indirect transition cannot be resolved whereas those for the y=0.045 and y=0.03 samples are clearly visible. The indirect gap energies are extracted from these features by fitting the signals and allowing corrections for residual strains and possible thermal shifts due to laser heating of the sample. The direct gaps peaks are also fitted using exponentially modified Gaussians (EMG) following procedures described in prior work. The Ge buffer layer PL peaks, also visible in some samples, are also accounted in our fits as shown in Figure 3 for a sample with y=0.045 . The data indicate that both the direct and indirect emission energies shift to lower values with the addition of Sn. The addition of Sn to the Ge lattice perturbs the pure Ge system such that with increased levels of Sn, both the Γ and L-point conduction band minima undergo a systematic reduction in energy relative to the Γ point valence band maximum, with the conduction band Γ point falling at a higher rate than the L-point. Thus at a critical Sn composition ($y=y_c$), the energy separation between the direct and indirect gaps vanishes and the material becomes a direct-gap semiconductor. Figure 4 shows linear fits to the direct and indirect gap energies extracted from PL spectra as a function of Sn composition indicating that the cross over composition (y_c) falls in the range of 8-10 %.

<u>Trigermane and tetragermane based $Ge_{1-y}Sn_y$ alloys</u>

From a historical perspective our prior work on ternary $Ge_{1-x-y}Si_xSn_y$ alloys had focused on materials with fixed Si/Sn ratios close to 4 (*x/y=4*), for which the crystals lattice-matched elemental Ge. These materials were grown on Ge substrates and did not exhibit PL in spite of the fact that diode structures fabricated from these layers showed defect densities lower than GeSn-on-Ge films, which do exhibit direct gap emission. The lack of emission from $Ge_{1-x-y}Si_xSn_y$ films lattice-matched to Ge is thus not due to inferior microstructure but a result of the increased separation between the direct and indirect edges relative to $Ge_{1-y}Sn_y$. The way to recover Ge-like emission properties from $Ge_{1-x-y}Si_xSn_y$ alloys is to produce Sn rich materials with $y>x$ concentrations, namely materials in which the direct and indirect edges are in close proximity to one another. In view of the relevance of $Ge_{1-x-y}Si_xSn_y$ alloys in Si photonic technologies outlined in the introduction, precise measurements of the lowest gap of the materials are required to enable mapping of the compositional dependence of the critical point energies allowing creation of device samples with efficient optical response by design. For this purpose a new series of $Ge_{1-x-y}Si_xSn_y$ was fabricated on Ge buffered Si(100) substrate via reactions of Ge_3H_8, Ge_4H_{10}, Si_4H_{10} and SnD_4 hydride precursors. These reacted at ultra-low temperatures 320-290°C to produce thick (~ 500 nm) monocrystalline films with a fixed 3-4 % Si content and progressively increasing Sn content in the 4-10 % range to explore the possibility of obtaining direct gap materials. The deposition and characterizations protocols in this case were similar to those described above for the UHV CVD depositions of $Ge_{1-y}Sn_y$ films on Ge/Si(100) substrates. The resultant films were found to be largely relaxed as evidenced by XRD analysis and they exhibit low defect densities and atomic scale chemical uniformity as indicated by STEM and element-selective EELS mapping, allowing a meaningful study of the optical properties as a function of concentrations.

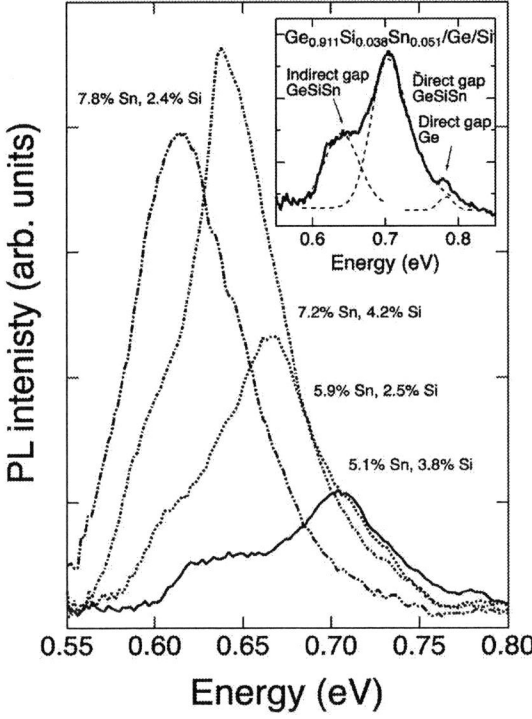

Figure 5: PL from $Ge_{1-x-y}Si_xSn_y$/Ge/Si samples, excited with 980 nm radiation. The main peak corresponds to direct-gap transitions, and the low-energy shoulder is assigned to the Ge-like indirect gap associated with the L-valley in the virtual crystal Brillouin zone. Inset shows fits of

the plots to extract the energy values of the various transitions including the direct gap of the Ge buffer layer (see Ref 51).

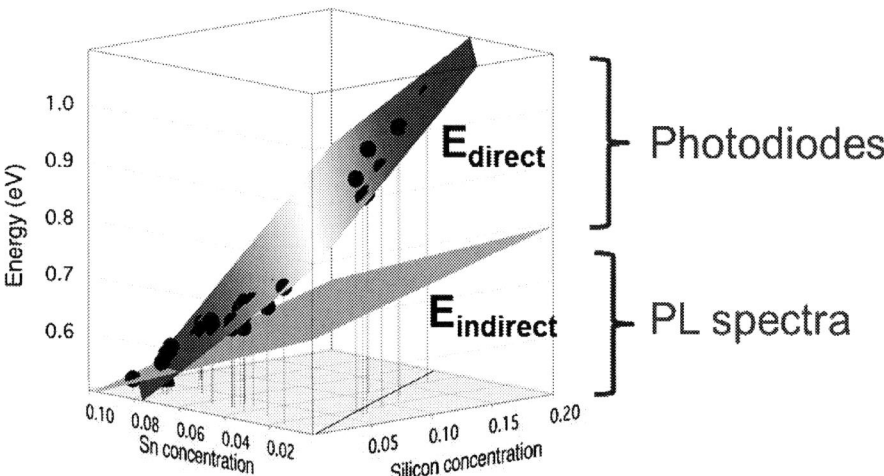

Figure 6: Summary of direct gap measurements for $Ge_{1-x-y}Si_xSn_y$ alloys combining of photocurrent data obtained from photodiodes exhibiting E_{direct} above that of Ge at 0.8 eV (spheres on blue plane) and photoluminescence data obtained from samples described in this paper (spheres on red plane). The Figure also includes the comparison of the measured indirect gaps (green lower plane) suggesting a cross-over from indirect to direct gap semiconductor for 9 % Sn and 3-4 % Si (see Ref 51).

The PL spectra of these films were measured and the optical emission was found to be strong enough to allow the determination of the direct gap and the indirect edge that represents the lowest band gap in the system over a broad compositional range. Figure 5 shows room temperature spectra from a series of samples, illustrating a dominant peak corresponding to direct-gap transitions and the weaker feature on the low-energy side corresponding to indirect gap transitions from the L minimum of the conduction band. The inset illustrates a fit of the data to extract the band gap values. A weak high-energy peak corresponding to direct-gap emission from the Ge buffer layer is fitted with a Gaussian. The results show that as the Sn- and Si-concentrations are adjusted in such a way that the direct gap decreases in energy, the separation between the direct and indirect gaps decreases and the photoluminescence intensity increases in these samples. The separation of the direct/indirect edges in this case can be made smaller than in Ge even for the non-negligible 3-4% Si content indicating that with a suitable choice of Sn compositions the ternary $Ge_{1-x-y}Si_xSn_y$ reproduces all features of the electronic structure of the binary $Ge_{1-y}Sn_y$, including the sought after indirect-to-direct gap cross. Figure 6 shows the compositional dependence of the measured E_0 direct gap energies for the $Ge_{1-x-y}Si_xSn_y$ alloys. The red plane spheres (located at lower portion of the top plane) correspond to the samples with compositions x>y described above in this paper. The blue plane spheres (located at upper portion of the top plane) are obtained from earlier responsivity measurement studies of samples with compositions $x/y = 4$ which are lattice matched to Ge.[48,51] The plot in the figure also compares the direct (red-white-blue top plane) and the indirect gaps (green bottom plane), indicating a crossover from indirect to direct gap semiconductor at the interception of the two planes corresponding to 9 % Sn and 3-4 % Si. The results described here represent the first demonstration of PL emission in $Ge_{1-x-y}Si_xSn_y$ semiconductors. The study culminates with the determination of the direct to indirect gap transition in this class of materials suggesting that the ternaries may offer an attractive alternative to $Ge_{1-y}Sn_y$ binary analogs for long wavelength applications in Si-based IR optoelectronic technologies.

CONCLUSION

This paper provides a detailed up-to-date account of research highlights illustrating accelerated progress in the ongoing development of SiGeSn semiconductor alloys, which have received global attention in recent years due to potential applications in Si-based photonic and microelectronic technologies. The paper focuses on the most recent fabrication and characterization at ASU of a new generation of optical quality binary and ternary alloys with Ge-rich compositions exhibiting direct gap behavior and strong light emission properties in the mid IR range of the spectrum. The photoluminescence results presented here indicate that the main features of the electronic structure of $Ge_{1-y}Sn_y$ films, (vanishing direct and indirect gap separation with increasing Sn content) can be reproduced in the $Ge_{1-x-y}Si_xSn_y$ ternary system. These observations make it possible to envision a wide range of optoelectronic applications including lasers, detectors and modulators with enhanced performance relative to Ge or SiGe, as well as high efficiency solar devices fully integrated on silicon.

ACKNOWLEDGMENTS

This work describe in this paper was partially supported by the U.S. Air Force under contract DOD AFOSR FA9550-12-1-0208.

REFERENCES

[1] R. F. C. Farrow, D. S. Robertson, G. M. Williams, A. G. Cullis, G. R. Jones, I. M. Young, and M. J. Dennis, J. Cryst. Growth **54**, 507 (1981).

[2] C. H. L. Goodman, Solid State and Electronic Devices, IEE Proceedings I **129** (5), 189 (1982).

[3] C. H. L. Goodman, Jap. J. of Appl. Phys. Supplement 22-1 **22** (Supplement 22-1), 583 (1982).

[4] S. Groves and W. Paul, Phys. Rev. Lett. **11**, 194 (1963).

[5] R. J. Temkin, G. A. N. Connell, and W. Paul, Solid State Commun. **11** (11), 1591 (1972).

[6] D. W. Jenkins and J. D. Dow, Phys. Rev. B **36** (15), 7994 (1987).

[7] K. A. Mäder, A. Baldereschi, and H. von Kanel, Solid State Commun. **69** (12), 1123 (1989).

[8] E. Kasper, J. Werner, M. Oehme, S. Escoubas, N. Burle, and J. Schulze, Thin Solid Films **520** (8), 3195 (2012).

[9] D. J. Paul, Semicond. Sci. Technol. **19** (10), R75 (2004).

[10] S. I. Shah, J. E. Greene, L. L. Abels, Q. Yao, and P. M. Raccah, J. Cryst. Growth **83**, 3 (1987).

[11] P. R. Pukite, A. Harwit, and S. S. Iyer, Appl. Phys. Lett. **54** (21), 2142 (1989).

[12] G. He and H. A. Atwater, Appl. Phys. Lett. **68** (5), 664 (1996).

[13] G. He and H. A. Atwater, Phys. Rev. Lett. **79** (10), 1937 (1997).

[14] V. R. D'Costa, C. S. Cook, A. G. Birdwell, C. L. Littler, M. Canonico, S. Zollner, J. Kouvetakis, and J. Menendez, Phys. Rev. B **73** (12), 125207 (2006).

[15] V. R. D'Costa, J. Tolle, R. Roucka, C. D. Poweleit, J. Kouvetakis, and J. Menendez, Solid State Commun. **144** (5-6), 240 (2007).

[16] R. Roucka, Y. Y. Fang, J. Kouvetakis, A. V. G. Chizmeshya, and J. Menéndez, Phys. Rev. B **81** (24), 245214 (2010).

[17] R. Beeler, R. Roucka, A. Chizmeshya, J. Kouvetakis, and J. Menéndez, Phys. Rev. B **84** (3), 035204 (2011).

[18] V. R. D'Costa, Y. Fang, J. Mathews, R. Roucka, J. Tolle, J. Menendez, and J. Kouvetakis, Semicond. Sci. Technol. **24** (11), 115006 (2009).

[19] A. V. G. Chizmeshya, C. Ritter, J. Tolle, C. Cook, J. Menendez, and J. Kouvetakis, Chem. Mater. **18** (26), 6266 (2006).

[20] Y. Y. Fang, J. Tolle, A. V. G. Chizmeshya, J. Kouvetakis, V. R. D'Costa, and J. Menendez, Appl. Phys. Lett. **95** (8), 081113 (2009).

[21] J. Q. Xie, J. Tolle, V. R. D'Costa, C. Weng, A. V. G. Chizmeshya, J. Menendez, and J. Kouvetakis, Solid-State Electronics **53** (8), 816 (2009).

[22] J. B. Tice, A. V. G. Chizmeshya, J. Tolle, V. R. D' Costa, J. Menendez, and J. Kouvetakis, Dalton Transactions **39** (19), 4551 (2010).

[23] J. Mathews, R. T. Beeler, J. Tolle, C. Xu, R. Roucka, J. Kouvetakis, and J. Menéndez, Appl. Phys. Lett. **97** (22), 221912 (2010).

[24] J. Mathews, R. Roucka, J. Q. Xie, S. Q. Yu, J. Menendez, and J. Kouvetakis, Appl. Phys. Lett. **95** (13), 133506 (2009).

[25] R. Roucka, J. Mathews, C. Weng, R. Beeler, J. Tolle, J. Menendez, and J. Kouvetakis, IEEE J. Quant. Electron. **47** (2), 213 (2011).

[26] R. Roucka, R. Beeler, J. Mathews, M.-Y. Ryu, Y. Kee Yeo, J. Menéndez, and J. Kouvetakis, J. Appl. Phys. **109** (10), 103115 (2011).

[27] R. Roucka, J. Mathews, R. T. Beeler, J. Tolle, J. Kouvetakis, and J. Menéndez, Appl. Phys. Lett. **98** (6), 061109 (2011).

[28] M. Bauer, C. Ritter, P. A. Crozier, J. Ren, J. Menéndez, G. Wolf, and J. Kouvetakis, Appl. Phys. Lett. **83** (11), 2163 (2003).

[29] Y.-Y. Fang, J. Xie, J. Tolle, R. Roucka, V. R. D'Costa, A. V. G. Chizmeshya, J. Menendez, and J. Kouvetakis, J. Am. Chem. Soc. **130** (47), 16095 (2008).

[30] J. Xie, A. V. G. Chizmeshya, J. Tolle, V. R. D'Costa, J. Menendez, and J. Kouvetakis, Chemistry of Materials **22** (12), 3779 (2010).

[31] V. R. D'Costa, Y. Y. Fang, J. Tolle, J. Kouvetakis, and J. Menéndez, Thin Solid Films **518** (9), 2531 (2010).

[32] V. R. D'Costa, Y. Y. Fang, J. Tolle, J. Kouvetakis, and J. Menendez, Phys. Rev. Lett. **102** (10), 107403 (2009).

[33] R. Roucka, J. Tolle, C. Cook, A. V. G. Chizmeshya, J. Kouvetakis, V. D'Costa, J. Menendez, Z. D. Chen, and S. Zollner, Appl. Phys. Lett. **86** (19), 191912 (2005).

[34] J. Tolle, R. Roucka, A. V. G. Chizmeshya, J. Kouvetakis, V. R. D'Costa, and J. Menendez, Appl. Phys. Lett. **88** (25), 252112 (2006).

[35] J. Menendez and J. Kouvetakis, Appl. Phys. Lett. **85** (7), 1175 (2004).

[36] J. Liu, X. Sun, R. Camacho-Aguilera, L. C. Kimerling, and J. Michel, Opt. Lett. **35** (5), 679 (2010).

[37] D. J. Friedman, S. R. Kurtz, and J. F. Geisz, presented at the Photovoltaic Specialists Conference, 2002. Conference Record of the Twenty-Ninth IEEE, 2002 (unpublished).

[38] G. Sun, H. H. Cheng, J. Menendez, J. B. Khurgin, and R. A. Soref, Appl. Phys. Lett. **90** (25), 251105 (2007).

[39] J. Faist, F. Capasso, D. L. Sivco, C. Sirtori, A. L. Hutchinson, and A. Y. Cho, Science **264** (22 April 1994), 553 (1994).

[40] W. Wang, S.-J. Su, J. Zheng, G.-Z. Zhang, Y.-H. Zuo, B.-W. Cheng, and Q.-M. Wang, Chinese Physics B **20** (6), 068103 (2011).

[41] S. Takeuchi, Y. Shimura, T. Nishimura, B. Vincent, G. Eneman, T. Clarysse, J. Demeulemeester, A. Vantomme, J. Dekoster, M. Caymax, R. Loo, A. Sakai, O. Nakatsuka, and S. Zaima, Solid-State Electronics **60** (1), 53 (2011).

[42] S. Su, B. Cheng, C. Xue, W. Wang, Q. Cao, H. Xue, W. Hu, G. Zhang, Y. Zuo, and Q. Wang, Opt. Express **19** (7), 6400 (2011).

[43] H. Lin, R. Chen, W. Lu, Y. Huo, T. I. Kamins, and J. S. Harris, Appl. Phys. Lett. **100** (10), 102109 (2012).

[44] A. Bhatia, W. M. Hlaing, G. Siegel, P. R. Stone, K. M. Yu, and M. A. Scarpulla, J. Elec. Mat. **41** (5), 837 (2012).

[45] S. Stefanov, J. C. Conde, A. Benedetti, C. Serra, J. Werner, M. Oehme, J. Schulze, D. Buca, B. Holländer, S. Mantl, and S. Chiussi, Appl. Phys. Lett. **100** (10), 104101 (2012).

[46] W.-J. Yin, X.-G. Gong, and S.-H. Wei, Phys. Rev. B **78** (16), 161203 (2008).

[47] G. Grzybowski, R. Roucka, J. Mathews, L. Jiang, R. Beeler, J. Kouvetakis, and J. Menéndez, Phys. Rev. B **84** (20), 205307 (2011).

[48] C. Xu, R.T. Beeler G. Grzybowski, A.V.G Chizmeshya J. Menendez and J. Kouvetakis *J. Am. Chem. Soc.* **134(51),** 20756-20767 (2012)

[49] G. Grzybowski, L. Jiang, R. T. Beeler, T. Watkins, A. V. G. Chizmeshya, C. Xu, J. Menéndez, and J. Kouvetakis, Chemistry of Materials **24** (9), 1619 (2012).

[50] C. Xu, R.T. Beeler, L. Jiang, A.V.G Chizmeshya J. Menendez and J. Kouvetakis, *Semicond. Sci. Technol.* **28,** 105001 (2013).

[51] J.D. Gallagher, C. Xu, L. Jiang, J. Kouvetakis, and J. Menéndez, *Appl. Phys. Lett.* **103,** 202104 (2013).

Mater. Res. Soc. Symp. Proc. Vol. 1666 © 2014 Materials Research Society
DOI: 10.1557/opl.2014.667

Effect of Annealing on Microstructure in (Doped and Undoped) Hydrogenated Amorphous Silicon Films

W. Beyer[1,2], W. Hilgers[2], D. Lennartz[2], F.C. Maier[2], N.H. Nickel[1], F. Pennartz[2], P. Prunici[3]
[1]Institut für Silizium-Photovoltaik, Helmholtz-Zentrum Berlin für Materialien und Energie, Kekuléstrasse 5, D-12489 Berlin, Germany
[2]IEK5-Photovoltaik, Forschungszentrum Jülich GmbH, D-52425 Jülich, Germany
[3]Malibu GmbH & Co.KG, Böttcherstrasse 7, D-33609, Bielefeld, Germany

ABSTRACT

Laser heating and annealing of hydrogenated amorphous silicon (a-Si:H) films is of interest for improved material properties. Due to the variety of possible laser treatments with regard to wavelength, pulse duration, scan time etc., the definition of laser impact on the material is a challenge which we try to approach by comparing properties of laser and oven treated materials. Here we report on the effect of oven heat treatment (up to $T_A= 575°C$) on microstructure and hydrogen content of hydrogenated amorphous silicon films, as detected by measurements of infrared absorption and of effusion of hydrogen as well as of implanted helium. The latter technique has been found to measure isolated voids (cavities) of the size of silicon divacancies and larger. Undoped as well as phosphorus and boron doped plasma-deposited a-Si:H films of various hydrogen content (< 15 at.%) were investigated, including undoped device grade a-Si:H. The results show little indication for void-related microstructure in the as-deposited and annealed state for material with a concentration of silicon bonded hydrogen below 5 at. %. At higher hydrogen concentration, evidence is found that hydrogen out-diffusion due to annealing causes isolated voids in concentrations up to about 10^{20} cm^{-3}. A possible mechanism for the annealing induced (micro-)void generation is discussed.

INTRODUCTION

Annealing effects in amorphous silicon (a-Si) are of interest for various reasons. On the one hand any deposition at elevated temperatures involves annealing effects of underlying layers or parts of the film. On the other hand, large area laser annealing may open possibilities for change and improvement of deposited films and devices, e.g. a-Si based solar cells. However, due to the variety of possible laser treatments with regard to wavelength, pulse duration, repetition rate, scan time etc., an improved understanding of laser annealing related changes in the material (along with a thorough characterization of the laser annealed state) are required. We try to approach this task by comparing laser-treated material with oven annealed material. We focus on annealing effects on void-related microstructure, which is an important property of a-Si materials. We characterize the microstructure by measurements of infrared absorption as well as effusion of implanted helium. Since helium does not react with the silicon network, helium effusion is known to be sensitive to microstructure and, in particular, isolated voids are detected since they trap diffusing helium [1]. Note, however, that only cavities equal or exceeding the size of divacancies trap helium in silicon [2, 3]. We limit the work to hydrogenated amorphous silicon (a-Si:H) material with hydrogen concentrations at or below 15 at.% H, i.e. to material which includes device-grade a-Si:H.

EXPERIMENT

The a-Si materials investigated involved hydrogenated amorphous silicon (a-Si:H) deposited at Research Center Jülich in a conventional rf plasma reactor with a substrate area of 50 cm^2. The silane flow was 3 sccm and an rf (13.56 MHz) power of 10 W was applied. Undiluted silane (gas pressure 0.5 mbar) was used as well as silane diluted 1:1 by a neon flow (gas pressure 1 mbar). The latter deposition conditions had the advantage of a higher film deposition rate and a (somewhat) reduced tendency for pinhole formation during deposition and during annealing [4]. For doping, flows of phosphine and diborane were added. Substrate temperature T_S ranged between 200 and 450°C. Also investigated was undoped a-Si:H of high substrate temperature ($T_S > 500$°C) and low H content deposited at Malibu Company, Bielefeld, using undiluted silane in a 10x10 cm^2 reactor as well as device-grade a-Si:H, fabricated at Research Center Jülich in a 40 x 40 cm^2 reactor using a gas mixture of 80 sccm SiH$_4$ and 330 sccm H$_2$ at 4 mbar, 33 W (13.56 MHz) rf power and a substrate temperature near 180°C.

The investigated films were typically 1 µm in thickness. Crystalline Si wafers were used as a substrate. In order to reduce pinhole formation during deposition or during annealing [4], the crystalline Si wafers were coated with thin layers of SiO$_2$ (typically 30 nm thickness) in some cases [5]. Annealing was done in vacuum in an UHV apparatus. Annealing time was 5 min. The samples were heated and cooled at high rates. Helium was implanted (using a mass separator) at the energy of 40 keV and at the (low) doses of 3x10^{15} cm^{-2} or 10^{16} cm^{-2}. The implantation energy corresponds to a mean helium depth of about 0.35 µm. Helium (and hydrogen) effusion measurements were performed as reported elsewhere [5] using a heating rate of 20°C per minute. By integrating over the low temperature LT (T< 600°C) and high temperature HT (T > 600°C) helium effusion, the ratio F* = HT/LT was determined and the void density N_V was estimated by the relation N_V = F* x 1.7 x10^{19} cm^{-3} [6]. Infrared absorption was measured using a Fourier transform infrared spectrometer. The integrated absorption in the range of the Si-H stretching modes was used to determine the density of bonded hydrogen [7]. To obtain the hydrogen concentration, a silicon atomic density of 5x10^{22} cm^{-3} was used. The infrared microstructure parameter R^{IR} [8] was determined from the ratio of integrated absorptions of the Si-H stretching band at 2100 cm^{-1} and of the total Si-H stretching absorption.

RESULTS

In Figs.1a-d and 2a-d the influence of annealing on effusion spectra of hydrogen and implanted helium is shown for undoped Si:H films deposited at $T_S = 200$°C (neon diluted silane) and $T_S = 400$°C (undiluted silane), respectively. In all cases, hydrogen shows a single relatively broad effusion maximum in the 550 – 700°C range indicating out-diffusion of H atoms from a dense material [5]. Analyzing the H effusion maxima in terms of H out-diffusion from a thin film [5], the effusion maxima of as-deposited $T_S = 400$°C and 200°C materials agree closely with previously reported H diffusion results [5] for high T_S and low T_S materials, respectively. For both materials the H effusion maxima are shifted slightly to higher temperatures with rising annealing temperature T_A. This shift can be explained by a decreasing H diffusion coefficient when H concentration is decreased in agreement with previous work [9, 10]. Helium effusion of the $T_S = 200$°C material (Fig. 1a) is seen to proceed in two stages. The low temperature He effusion near 400°C has been attributed to He diffusion through the bulk Si:H material while He effusion at temperatures exceeding about 600°C is attributed to He trapped in voids [1]. The

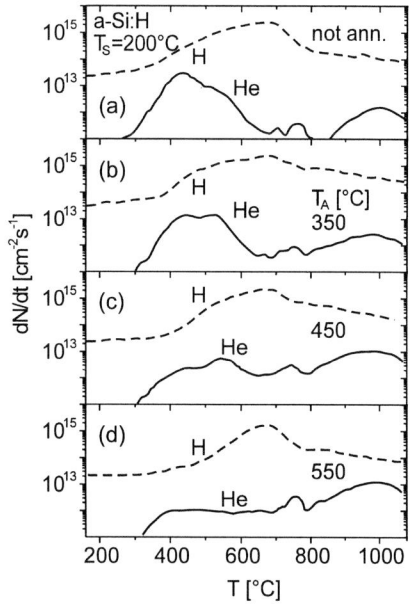

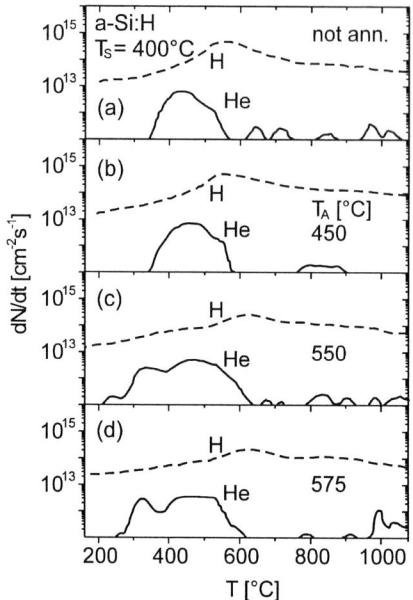

Figure 1. Effusion rate of H and implanted He (dose: 10^{16} cm^{-2}) for a-Si:H film deposited at $T_S = 200°C$, not annealed (a) and annealed at various temperatures T_A ((b)-(d)).

Figure 2. Effusion rate of H and implanted He (dose: 3×10^{15} cm^{-2}) for a-Si:H film deposited at $T_S = 400°C$, not annealed (a) and annealed at various temperatures T_A ((b)-(d)).

small structure near 800°C is attributed to film crystallization [5]. It is seen that annealing leads to a relative increase of high temperature He effusion in (qualitative) agreement with previous work [1, 11]. The $T_S = 400°C$ material in Fig. 2, on the other hand, shows no significant high temperature He effusion, both in the not annealed and the annealed state. Note that dN/dt signals near 10^{11} cm^{-2} s^{-1} are close to the background/ measuring limit of the gas effusion system. For this $T_S = 400°C$ material, however, an additional He effusion structure near 330-350°C appears at annealing temperatures exceeding 450°C. Since some bubble and blister formation [4] was observed for this sample at $T_A > 450°C$, we attribute this helium peak to implanted He trapped in bubbles and/or to He implanted through bubbles or pinholes into the c-Si wafer material.

In Fig. 3, for the two a-Si:H samples of Figs. 1 and 2 the results of measurements of He effusion and of infrared absorption as a function of annealing temperature T_A are compiled. Plotted are the concentration C_H of bonded hydrogen (Fig. 3a), the infrared microstructure parameter R^{IR} (Fig. 3b) and the estimated density of voids (trapping helium) N_V (Fig. 3c) as a function of annealing temperature T_A. Also shown are the results of an undoped a-Si:H film deposited under "device grade" conditions (see experimental section) with $C_H \approx 12$ at.%. Both $C_H > 10$ at. % samples show a quite similar behaviour: R^{IR} rises slightly whereas N_V increases rather pronounced as T_A increases to 550°C. This latter increase is particularly strong when the H concentration shows the strongest decrease, i.e for the $T_S = 200°C$ sample at $T_A > 350°C$ and for the device-grade sample at $T_A > 450°C$ (see Fig. 3a). Roughly one void is formed when for the $T_S = 200°C$ material and the device-grade material 40 and 80 H atoms effuse out, respectively. In contrast, the $T_S = 400°C$ material shows $R^{IR} \approx 0$ and N_V near 10^{18} cm^{-3}, independent of the annealing state, although more than 2 at.% (10^{21} cm^{-3}) of bonded H diffuses out. We note that

38

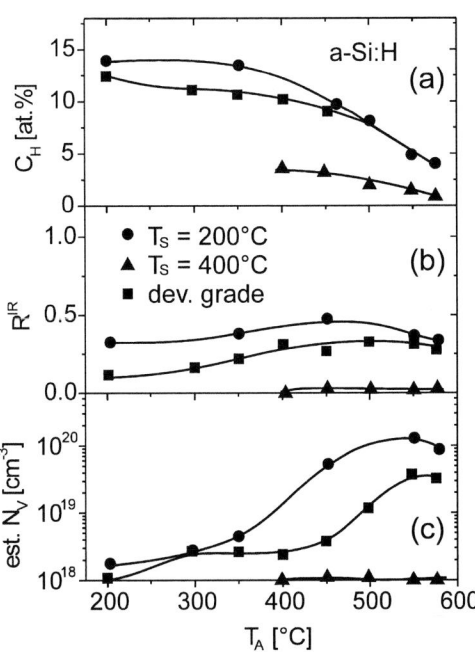

Figure 3. Concentration of silicon-bonded hydrogen C_H (a), infrared microstructure parameter R^{IR} (b) and estimated void density N_V (c) as a function of annealing temperature T_A (annealing time 5 min.) for a-Si:H deposited at $T_S = 200°C$, $400°C$ and for device-grade a-Si:H.

these latter results agree closely with previous data of H implanted material of similar (3 at. %) hydrogen concentration [12].

Exploring these differences in the annealing behaviour further, we plot in Fig. 4 R^{IR} and N_V data versus hydrogen concentration C_H (measured prior to annealing) of as-deposited materials as well as for the strongest deviation from the as-deposited state, usually reached after 550°C annealing. The data include the samples of Fig. 3. It is seen that for $C_H < 4$-5 at.%, R^{IR} and N_V are small both in the as-deposited and annealed state. At $C_H > 4$-5 at.%, both R^{IR} and N_V increase steadily with rising H concentration in the annealed state while in the as-deposited state both parameters (R^{IR} and N_V) remain rather low except for R^{IR} of the material with the highest hydrogen concentration ($T_S = 200°C$).

For phosphorus-doped a-Si:H, the dependence of R^{IR} and N_V on C_H is similar to undoped material in the annealed state. In the as-deposited state, however, the values of R^{IR} and N_V are small only for material with $C_H < 5$ at.%. At higher C_H, both N_V and R^{IR} attain rather high values in the as-deposited state, too.

For boron-doped material, more work appears necessary. The present results show for R^{IR} in the whole H concentration range and for N_V at $C_H < 5$ at.% similar results as observed for undoped material. But for B-doped material of $C_H > 5$ at.%, a steep increase of N_V to high values of $N_V \approx 10^{20}$ cm^{-3} is found both for the as-deposited and annealed state. It must be noted, however, that due to H effusion from B-doped material at rather low temperatures [5] it cannot be excluded that the implanted helium detects (in part) voids which are formed by H out-diffusion during the effusion experiment.

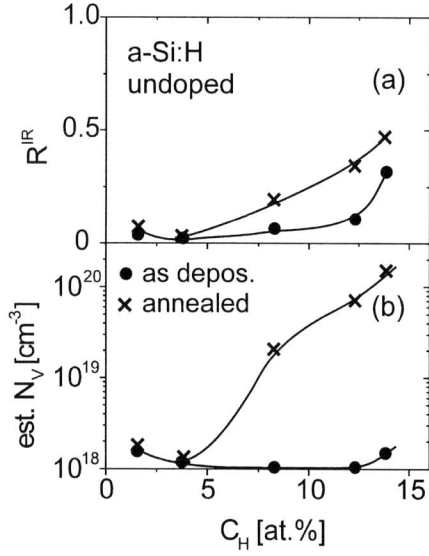

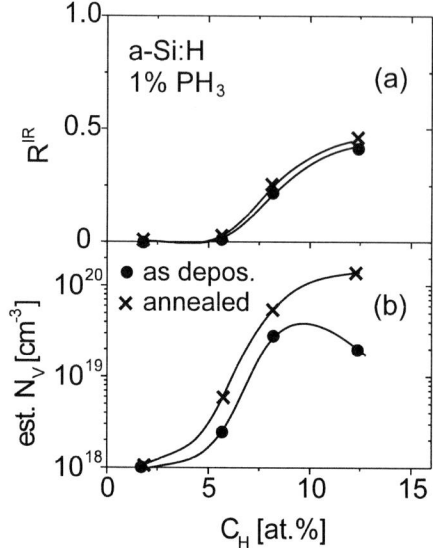

Figure 4. Infrared microstructure parameter R^{IR} (a) and estimated void density N_V (b) for undoped a-Si:H (as deposited and after annealing) versus concentration C_H of silicon-bonded hydrogen.

Figure 5. Infrared microstructure parameter R^{IR} (a) and estimated void density N_V (b) for phosphorus doped a-Si:H (as deposited and after annealing) versus concentration C_H of silicon-bonded hydrogen.

DISCUSSION

A main result of the present work is the evidence for the formation of voids of the size of divacancies and larger by H out-diffusion. This refers to a-Si:H with hydrogen concentrations exceeding about 4-5 at.% and includes device-grade a-Si:H. Similar increases in (He-trapping) void concentration upon annealing, observed for rather low substrate temperature (non device-grade) material were tentatively attributed to a conversion of interconnected voids to isolated ones [1,11]. However, for the present set of samples there is no indication for the presence of interconnected voids which should reveal in a low-temperature hydrogen effusion peak [5]. Since material density in a-Si:H is known to be reduced by hydrogen incorporation [17], it appears likely that H out-diffusion leaves behind cavities large enough for the trapping of diffusing He, if sufficient H has effused out. Out-diffusion of H is known to result in Si dangling bonds or weak Si-Si bonds. Clustering (agglomeration) of such bonds may result in void formation although the material will predominantly contract [5]. This can explain the proportionality of N_V and effusing H. We note that voids including the isolated voids detected by He out-diffusion likely cause the time-dependence of the hydrogen diffusion coefficient [9, 14, 15], as diffusing H may precipitate as H_2 in these voids. A strong increase of time dependence of the H diffusion coefficient by annealing has been reported for H rich a-Si:H [16]. Clustering of weak or broken Si-Si bonds (i.e. void formation) may not occur for $C_H < 5$ at.% material. Since the hydrogen (Si-H) bonding site is considered to involve a cavity size of about a monovacancy [18], isolated Si-H bonding sites cannot be expected to trap helium [2, 3]. This agrees with the finding [12] that a-Si implanted with about 3 at.% (1.5×10^{21} cm^{-3}) of hydrogen (which showed up in infrared absorption to be in bonds with silicon) has an estimated density of voids trapping helium of only 10^{18} cm^{-3}. With regard to an aimed improvement of a-Si:H films by annealing, it seems clear that such void formation needs to be kept small, either by minimizing H out-diffusion or by employing films of low hydrogen concentration. It is also conceivable that improved deposition techniques can reduce the hydrogen out-diffusion related void formation.

CONCLUSIONS

Microstructure characterization by infrared absorption and by effusion measurements of implanted helium shows for a-Si:H with a hydrogen concentration exceeding about 5 at.% that voids trapping He are generated when silicon-bonded hydrogen effuses. Accordingly, for improved a-Si:H material properties by annealing, such void formation needs to be minimized.

ACKNOWLEDGEMENTS

The authors wish to thank Dr. T. Merdzhanova and her group for film depositions and A. Dahmen for the ion implantations. Part of the work was financed by BMU ("Globe-Si" project No. 0325446A).

REFERENCES

1. W. Beyer, Phys. Status Solidi (c) **1**, 1144 (2004).
2. S.K. Estreicher, J. Weber, A. Derecskei-Kovacs and D.S. Marynick, Phys.Rev. B **55**, 5037 (1997).
3. A.A. Gnidenko, V.G. Zavodinsky, A. Misiuk, J. Bak-Misiuk, Acta Physica Polonica A **109**, 353 (2006).
4. H.R. Shanks and L. Ley, J. Appl. Phys. **52**, 811 (1981).
5. W. Beyer, F. Einsele, in Advanced Characterization Techniques for Thin Film Solar Cells, edited by D. Abou-Ras, T. Kirchartz, U. Rau (Wiley-VCH, Weinheim, Germany, 2011) p. 449.
6. W. Beyer, W. Hilgers, P. Prunici, D. Lennartz, J. Non-Cryst. Solids **358**, 2023 (2012).
7. W. Beyer and M.S. Abo Ghazala, MRS Symp. Proc. **507**, 601 (1998).
8. A.H. Mahan, P. Raboisson, D.I. Williamson, R. Tsu, Solar Cells **21**, 117 (1987).
9. W. Beyer, in: Semiconductors and Semimetals 61, N.H. Nickel, ed. (Academic Press, San Diego, 1999) p. 154.
10. W. Beyer, Solar Energy Materials and Solar Cells **78**, 235 (2003).
11. W. Beyer, D. Lennartz, P. Prunici, H. Stiebig, MRS Symp. Proc. **1321**, 135 (2011).
12. W. Beyer, U. Breuer, R. Carius, W. Hilgers, D. Lennartz, F.C. Maier, N.H. Nickel, F. Pennartz, P. Prunici and U. Zastrow, Can. J. Phys. **92** (2014) in print.
13. W. Beyer, W. Hilgers, D. Lennartz, F.C. Maier, N.H. Nickel, F. Pennartz, P. Prunici, MRS Symp. Proc. **1536**, 175 (2013).
14. R.A. Street, C.C. Tsai, J. Kakalios, W.B. Jackson, Philos. Mag. B **56**, 305 (1987).
15. X.M. Tang, J. Weber, Y. Baer, F. Finger, Phys. Rev. B **41**, 7945 (1990).
16. X. M. Tang, J. Weber, Y. Baer, F. Finger, Phys. Rev. B **42**, 7277 (1990).
17. Z. Remes, M. Vanecek, P. Torres, U. Kroll, A.H. Mahan, R.S. Crandall, J. Non-Cryst. Solids **227-230**, 876 (1998).
18. M. Cardona, Phys. Status Solidi (b) **118**, 463 (1983).

Mater. Res. Soc. Symp. Proc. Vol. 1666 © 2014 Materials Research Society
DOI: 10.1557/opl.2014.683

Hydrogenated amorphous silicon germanium by Hot Wire CVD as an alternative for microcrystalline silicon in tandem and triple junction solar cells

L.W. Veldhuizen[1], Y. Kuang[2], N.J. Bakker[3], C.H.M. van der Werf[3], S.-J. Yun[4], R.E.I. Schropp[1,3]

[1]Eindhoven University of Technology (TU/e), Department of Applied Physics, Plasma & Materials Processing, P.O. Box 513, 5600 MB Eindhoven, The Netherlands
[2]Physics of Devices, Debye Institute for Nanomaterials Science, Utrecht University, High Tech Campus 2, 5656 AE Eindhoven, The Netherlands
[3]Energy research Center of the Netherlands (ECN), ECN-Solliance, High Tech Campus Building 2, 5656 AE Eindhoven, The Netherlands
[4]Thin Film Solar Cell Technology Team, Convergence Components and Materials Research Laboratory, Electronics and Telecommunications Research Institute, 218 Gajeongno, Yuseong-gu, Daejeon 305-700, Republic of Korea

ABSTRACT

We study hydrogenated amorphous silicon germanium (a-SiGe:H) deposited by HWCVD for the use as low band gap absorber in multijunction junction solar cells. We deposited layers with Tauc optical band gaps of 1.21 to 1.56 eV and studied the hydrogen bonding with FTIR for layers that were deposited at several reaction pressures. For our reaction conditions, we found an optimal reaction pressure of 38 µbar. The material that is obtained under these conditions does not meet all device quality requirements for a-SiGe:H, which is, as we hypothesize, caused by the presence of He that is used to dilute the GeH_4 source gas. We present an initial single junction n-i-p solar cell with a Tauc optical band gap of 1.45 eV and a short circuit current density of 18.7 mA/cm^2.

INTRODUCTION

The manufacturing cost of thin film Si based tandem and triple junction cells and modules is at present too high for thin film Si modules to meet current module market prices. Conventionally, microcrystalline silicon is used as the low band gap absorber in 'micromorph' solar cells (a-Si/µc-Si tandem cells). However, due to the considerable thickness needed for the µc-Si:H absorber, it takes three to four times as many deposition reactors compared to single junction cells to produce tandem cells, leading to high cost of ownership. One of the approaches to reduce processing time of the low band gap layer(s) in multijunction silicon-based solar cells is the use of hydrogenated amorphous silicon germanium (a-SiGe:H). Until recently, United Solar Ovonic has been very successful in the development of triple junction solar cells with a-SiGe:H absorber layers [1]. In general however, a-SiGe:H has not been regarded as a viable option because of (i) the high defect density for PECVD a-SiGe:H, at band gaps < 1.4 eV, and [2] (ii) the high cost of germane (GeH_4). On the other hand, because of its direct gap nature, the thickness of an a-SiGe:H absorber layer can be kept 10 times smaller than that of µc-Si:H. The use of GeH_4 is a viable option if cells can be made ultra-thin, by the implementation of light scattering nanostructure.

We are investigating whether a-SiGe:H can be (re-)considered for inexpensive production of multijunction thin film Si based solar cells if Hot-wire CVD (HWCVD) is used as the deposition method. HWCVD is a simple and low cost deposition technique allowing high deposition rates while maintaining good defect passivation. Early results reported by NREL include the achievement of material with a band gap close to that of µc-Si:H (1.2 eV) with an equivalent photoresponse (in excess of two orders of magnitude) [3]. Their work has led to 8.64% single junction cells, obtained without any band gap profiling in the absorber layer [4]. We now continued this development to provide a novel thin film alloy for the struggling 'micromorph' thin film silicon technology.

EXPERIMENTAL DETAILS

a-SiGe:H films with thicknesses of ~200 nm were deposited by HWCVD, using two 0.5 mm diameter Ta filaments that each have a length of 15 cm. The filaments were mounted parallel to each other and were separated by a distance of 4 cm. Corning glass and natively oxidized polished c-Si wafers were simultaneously used as substrates and were located, facing down, 5.5 cm above the filaments. The background pressure in the reactor was $<10^{-7}$ mbar.

During the deposition, the filaments had a temperature around 1700°C. The substrates were heated by radiation from the filaments and additional heating could be provided by an external heater. As source gasses, high purity SiH_4 and GeH_4 were used, the latter of which is diluted to 9.8% in He for safety considerations. Additionally, H_2 was added to the process gas mixture since it was reported that H_2 dilution can improve the quality of the film due to an increase in H coverage of the growing surface, selective etching of Ge-Ge weak bonds, and the production of a more homogeneous structure [5,6,7].

The complex refractive index as well as the thickness of the films were determined from reflection and transmission measurements in combination with the model of O'Leary, Johnson, and Lim (OJL) [8]. The complex refractive index was used to calculate the optical band gap of the materials with two methods. The Tauc band gap (E_{Tauc}) is found by extrapolating the linear regime of $(\alpha nE)^{1/2}$ vs. E to $\alpha nE = 0$ and a more unambiguously defined optical gap $E_{3.5}$ is defined as the energy at which the absorption coefficient α exceeds the value of $10^{3.5}$ cm^{-1}[9].

The Ge film content $C_{Ge, XPS}$ was determined by measuring the contributions of the Ge 3d and Si 2p peaks for several sample depths with X-ray photoelectron spectroscopy (XPS) and Al Kα radiation. Additionally, we estimate the Ge content in the films by comparing the peak areas of the amorphous transverse optical modes (TO) of Ge-Ge (270 cm^{-1}) and Si-Ge (375 cm^{-1}) as measured by Raman spectroscopy, using a 514.5 nm Ar$^+$ laser. Assuming a randomly distributed network of Si and Ge atoms, $C_{Ge, Raman}$ can be expressed as [5]:

$$C_{Ge, Raman} = \frac{2\,I_{Ge-Ge}}{2\,I_{Ge-Ge} + I_{Si-Ge}} \cdot 100\% \qquad (1)$$

Fourier transformed infrared spectroscopy (FTIR) with a liquid nitrogen cooled MCT detector was applied to analyze the bonding concentrations and configurations of H to Si and Ge in the region of 375-4000 cm^{-1}. The frequency dependent absorption coefficient of the films was calculated by measuring the IR transmission of the films that were deposited on polished c-Si wafers. The absorption coefficient was corrected for incoherent as well as coherent multiple reflections [10,11]. The absorption profiles were deconvoluted with Gaussian profiles from which the bonding concentration N_k is calculated as:

$$N_k = A_k I_k = A_k \int \frac{\alpha_k(\omega)}{\omega} d\omega \qquad (2)$$

where A_k and I_k are the proportionality constant and the integrated absorption for the vibrational mode k respectively.

For the electrical measurements, coplanar Ag contacts were evaporated on the films. The dark conductivity σ_d and the activation energy (E_A) of the dark conductivity were measured in vacuum at a voltage of 50 V after annealing the films at 160°C for 90 minutes. The photo-conductivity σ_p was measured at atmospheric pressure under AM1.5 illumination.

RESULTS AND DISCUSSION

The effect of the Ge content

We deposited several films with different Ge concentrations using a flow of 15 sccm GeH_4 while choosing a SiH_4 flow varying from 0 to 22.5 sccm. The gas mixture was diluted with 100 sccm H_2. The reaction pressure was kept constant at 50 µbar. No external heating was applied, which resulted in a substrate temperature of 200 °C during deposition. Table I shows the material properties of a series of materials with a varying GeH_4 gas flow ratio $X_{GeH_4} = [GeH_4/ (GeH_4 + SiH_4)]$.

Table I. Material properties of films with varying X_{GeH_4}.

X_{GeH_4}	R_d nm/s	$C_{Ge, XPS}$ %	$C_{Ge, Raman}$ %	$E_{3.5}$ eV	E_{Tauc} eV	E_a eV	σ_d S/cm	σ_p S/cm	σ_p/σ_d
0.40	0.74	56.6	56.1	1.52	1.56	0.91	9.54E-12	7.69E-08	8.06E+03
0.46	0.63	62.7	64.3	1.46	1.51	0.88	3.10E-11	9.21E-08	2.97E+03
0.55	0.63	70.7	70.9	1.40	1.45	0.81	1.43E-10	1.08E-07	7.52E+02
0.67	0.56	79.7	78.6	1.31	1.35	0.68	3.38E-09	4.14E-07	1.23E+02
0.86	0.52	93.7	96.4	1.17	1.22	0.57	5.01E-07	1.01E-05	2.02E+01
1.00	0.71	99.3	97.1	1.13	1.21	0.43	3.15E-06	1.09E-05	3.45E+00

The deposition rate R_d of the films in this series is always high, in excess of 0.5 nm/s. With this deposition rate, an active absorber layer (i-layer) of 150 nm will readily be deposited within 5 minutes. The Ge concentration in the films is higher than the ratio X_{GeH_4} in the reactive gas mixture, showing that Ge is preferential incorporated. This preferential incorporation of Ge was also observed in previous studies and can be explained by both the higher dissociation rate of GeH_4 as compared to SiH_4, due to the weaker bond strength of Ge-H compared Si-H bonds, and the lower mobility of Ge-H_x radicals compared to Si-H_x radicals [3,12].

In previous research, clustering of Ge atoms has been observed in a-SiGe:H films [5]. To investigate the amount of Ge clustering we compared the Ge content as determined by XPS with the Ge content as determined with Raman spectroscopy using equation 1. As can be seen in table I, there is a good agreement between $C_{Ge, XPS}$ and $C_{Ge, Raman}$. We therefore conclude that there is no significant clustering in these materials.

Table I also shows the optical band gap, the activation energies and the corresponding room temperature dark conductivities of the materials. E_{Tauc} varied between 1.56 and 1.21 eV for materials with Ge concentrations between 56.6−99.3%. For these Ge concentrations, the optical band gaps are up to 0.2 eV higher than the band gaps that were reported in previous research [3]. The increased band gaps are an indication that the materials have a high microvoid density. In the next section we will therefore have a closer look at the material structure.

The influence of the reaction pressure

In this section we focus on materials that have a Ge concentration of ~70% and have a Tauc band gap of ~1.45 eV. We studied the effect of varying the reaction pressure from 12 to

80 µbar on the material structure . In order to cover this pressure range, we used a reduced flow of 6.4 sccm SiH$_4$, 7.5 sccm GeH$_4$ and 50 sccm H$_2$. We applied moderate external heating, resulting in a substrate temperature of 246 °C, since we found that this was beneficial for the material properties during initial optimization of the process conditions. Using FTIR we analyzed the absorbance peaks around 580 cm^{-1} and 635 cm^{-1} that are associated with the wagging modes of hydrides of Ge and Si, respectively. We calculate the atomic hydrogen concentration C$_H$ by multiplying the absorption coefficient intensity I$_{580}$ and I$_{635}$ with the proportionality constants A$_{580}$ = 1.1·10^{19} cm^{-3} and A$_{635}$ = 2.1·10^{19} cm^{-3} respectively as in equation 2, and dividing their sum by the atomic density of the material which was estimated to be 5· (1− C$_{Ge}$/100) + 4.4· C$_{Ge}$/100)·10^{22} cm^{-3} [11,13,14].

We also analyzed the material structure by measuring the intensity of the stretching mode at 2000 cm^{-1}, which can be associated to Si monohydrides and the stretching mode around 2090 cm^{-1} which represents Si dihydrides or monohydrides at internal surfaces of voids [15]. The stretching modes related to Ge mono- and dihydrides are positioned at 1880 cm^{-1} and 1980 cm^{-1}, respectively, the latter of which cannot be distinguished from the Si monohydrides. Since we did not find any absorption at this position in the material with C$_{Ge}$= 99.3%, we are confident to neglect this mode. Using the absorption intensities from the stretching modes, we calculate the microstructure parameter R*:

$$R^* = \frac{I_{2090}}{I_{2000} + I_{2090}} \tag{3}$$

A high microstructure parameter is often related to a high microvoid density and therefore suggests a poor material quality [16, 17].

Another indication of the material quality is the preferential attachment of H to Si over Ge, P$_{Si}$ that is defined as [18]:

$$P_{Si} = \frac{I_{2000}}{I_{1880}} \cdot \frac{C_{Ge}}{100 - C_{Ge}} \tag{4}$$

High values of P$_{Si}$ have been related to high densities of Ge dangling bonds [18]. Lower values of R* and P$_{Si}$ have indeed been reported to result in higher efficiency a-SiGe:H solar cells [4].

Figure 1a shows the IR absorption coefficient of the Si hydride and Ge hydride stretching modes for different reaction pressures. The trends of the IR absorption have been quantified in figure 1b, in which R*, P$_{Si}$ and C$_H$ are shown as function of the reaction pressure. The steep decline of R* and P$_{Si}$ in the low pressure region can be explained by considering that for a pressure of 12 µbar, the mean free path of the radicals in the reactor is in the order of the filament to substrate distance. This means that atomic Si and Ge, which is produced at the filament surface, and radicals with low hydrogen content are the main growth precursors. Because of the high sticking coefficient of these radicals, the film will grow with a rich microstructure, explaining the high R* at low pressures. Additionally, the low amount of available hydrogen at the surface will preferentially react with Si because of the higher binding energy of Si-H as compared to Ge-H, causing high P$_{Si}$ values. An optimum is observed at 38 µbar after which R* slowly increases again. Two possible reasons for the increase of the microstructure parameter at higher pressures are an increase of deposition rate and the formation of higher silanes in the gas phase [19]. The total hydrogen concentration C$_H$ for all the films in this series is around 9%.

The values of R* and P$_{Si}$ that we observed for the growth at 38 µbar are higher than previously reported device-quality a-SiGe:H [4]. We hypothesize that the high flows of He, which are used to dilute GeH$_4$ in the gas cylinder, and consequently, in our reactor, contributes

45

to the inferior material quality. Although He is an inert gas, we cannot exclude that He alters the growth process by occupying growth sites or cooling the filament and the substrate.

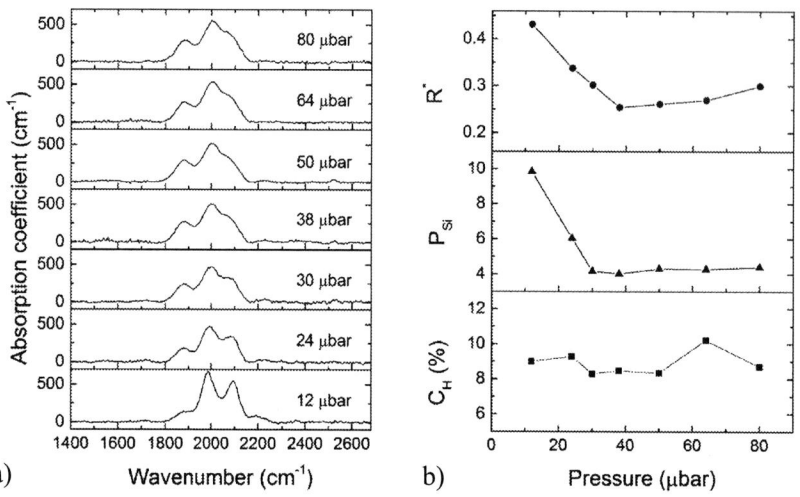

Figure 1. a) IR absorption coefficient of the Si hydrides and Ge hydrides stretching modes for a series of reaction pressures. b) Microstructure parameter R* (●), preferential attachment factor P_{Si} (▲) and hydrogen concentration C_H (■) as a function of the reaction pressure.

Initial cell results

We have tested the material that is deposited using the optimized pressure conditions in a single junction a-SiGe:H solar cell in substrate configuration (n-i-p) on Asahi-U glass that is coated with 200 nm Ag and 100 nm ZnO:Al. 80 nm ITO was used as front TCO. We present the internal collection efficiency (ICE) of our initial device in figure 2.

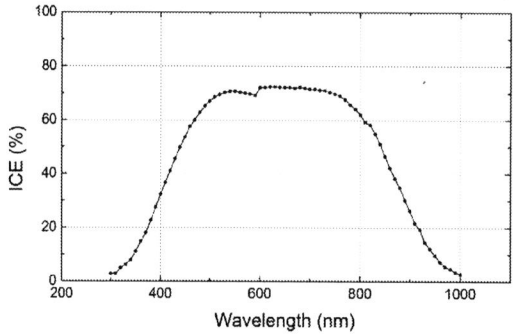

Figure 2. Internal collection efficiency (ICE) of an a-SiGe:H n-i-p solar cell.

The ICE shows absorption in a broad range of wavelengths, which can be explained by the low optical band gap of the intrinsic material (E_{Tauc} = 1.45 eV). Despite an intrinsic layer thickness of only 100 nm, the short circuit current density (J_{sc}) of this cell equals 18.7 mA/cm^2. In order to incorporate the intrinsic a-SiGe:H into a solar cell effectively, its interfaces and band gap profiling have to be carefully optimized. We expect that the generated current of this type of solar cell can be increased by further optimizing the material quality of the intrinsic a-SiGe:H and introducing appropriate band gap profiling.

CONCLUSIONS

a-SiGe:H films with varying Ge concentrations were deposited by HWCVD resulting in Tauc optical band gaps of 1.21 to 1.56 eV. The microstructure and preferential attachment of H to Si were studied with FTIR for layers that were deposited at several reaction pressures. For our reaction conditions, with filament distance of 5.5 cm, we found an optimal reaction pressure of 38 μbar. The material that is obtained under these conditions is not yet optimal, which is possibly caused by presence of He that is used to dilute the GeH_4 source gas. Nevertheless, we have incorporated the present best material with a Tauc optical band gap of 1.45 eV into an initial single junction n-i-p solar cell, resulting in a short circuit current density of 18.7 mA/cm^2.

ACKNOWLEDGMENTS

This work is supported by the Energy International Collaboration Project of the Korea Institute of Energy Technology Evaluation and Planning (KETEP) and NanoNextNL, a micro and nanotechnology consortium of the Government of the Netherlands and 130 partners.

REFERENCES

1. S. Guha, D. Cohen, E. Schiff, P. Stradins, P.C. Taylor, J. Yang, *Photovoltaics International* **13**, 134 (2011).
2. A. Matsuda, K. Yagii, M. Koyama, M.Toyama, Y.Imanishi, N. Ikuchi, and K.Tanaka, *Appl. Phys. Lett.* **47**, 1061 (1985).
3. Y. Xu, A.H. Mahan, L.M. Gedvilas, R.C. Reedy, H.M. Branz, *Thin Solid Films* **501**, 198 (2006).
4. A.H. Mahan, Y. Xu, L.M. Gedvilas, D.L. Williamson, *Thin Solid Films* **517**, 3532 (2009).
5. J. Yang, L. Newton, and B. Fieselmann, in *Amorphous and Heterogeneous Silicon-Based Films* (Mater. Res. Soc. Symp. Proc. **149**, Warrendale, PA, 1989) p. 497.
6. M. Shima, M. Isomura, E. Maruyama, S. Okamoto, H. Haku, K. Wakisaka, S. Kiyama, and S. Tsuda, Jpn. *J. Appl. Phys.* **37**, 6322 A (1998).
7. A. Catalano, R. Arya, M. Bennett, L. Yang, J. Morris, B. Goldstein, B. Fieselmann, J. Newton, and S. Wiedeman, *Sol. Cells* **27**, 25 (1989).
8. S. O'Leary, S. Johnson, and P. Lim, *J. Appl. Phys.* **82**, 3334 (1997).
9. J. Tauc, in *Amorphous and liquid semiconductors* (Plenum Press, London, 1974) p. 178.
10. M.H. Brodsky, M. Cardona, J.J. Cuomo, *Phys. Rev. B* **16**, 3556 (1977).
11. A.A. Langford, M.L. Fleet, B.P. Nelson, W.A. Lanford, N. Maley, *Phys. Rev. B* **45**, 13367 (1992).
12. Y.P. Chou, and S.C. Lee, *J. Appl. Phys.* **83**, 4111 (1998).
13. C.J. Fang, K. J. Gruntz, L. Ley, M. Cardona, F.J. Demond, G. Müller, S. Kalbizer, *J. Non-Cryst. Solids* **35-36**, 255 (1980).
14. A. Morimoto, T. Miura, M. Kumeda, and T. Shimizu, *Jpn. J. Appl. Phys.* **20**, L833 (1989).
15. J. Daey Ouwens, R.E.I. Schropp, and W. van der Werf, *Appl. Phys. Let.* **65**, 204 (1994).
16. A.H. Mahan, P. Raboisson, R. Tsu, *Appl. Phys. Lett.* **50**, 335 (1987).
17. D.E. Carlson, in *Materials Issues in Amorphous Semiconductor Technology*, edited by D. Adler, Y. Hamakawa, and A. Madan, (Mat. Res. Soc. Sym. Proc. 70 (Mater. Res. Soc. Symp. Proc. **70**, Pittsburgh, PA, 1986), p. 467.
18. W. Paul, D.K. Paul, B. von Roedern, J. Blake, S. Oguz, *Phys. Rev. Lett.* **46**, 1016 (1981).
19. E.C. Molenbroek, A.H. Mahan, A. Gallagher, *J. Appl. Phys.* **82**, 1909 (1997).

Mater. Res. Soc. Symp. Proc. Vol. 1666 © 2014 Materials Research Society
DOI: 10.1557/opl.2014.716

High quality kerfless silicon mono-crystalline wafers and cells by high throughput epitaxial growth

R. Hao, T.S. Ravi, V. Siva, J. Vatus, D. Miller, J. Custodio, K. Moyers
Crystal Solar Inc, 3050 Coronado Drive, Santa Clara, CA 95054, U.S.A.

ABSTRACT

Crystalline silicon based photovoltaics continues to be the dominant technology for large scale deployment of solar energy. While impressive cost gains in silicon based PV have come with scale, there remains a strong push for increased efficiencies and further lowering of manufacturing costs to achieve true grid parity. So far, however, there has not been a production proven approach that reduces the cost while maintaining or increasing the efficiency. Attempts to reduce the amount of silicon used, for example, have led to development of various kerfless wafer manufacturing approaches. While some of these approaches have shown the potential for reduced costs, they also compromise the efficiency mainly due to the inferior quality of the material.

Epitaxy based kerfless silicon wafers, on the other hand, has shown the potential to reverse this trend offering lower manufacturing costs while maintaining or even enhancing the efficiency due to the high quality of the n-type and p-type silicon epitaxial (Epi) wafers. In this work, we present key aspects of Crystal Solar's patented high throughput production silicon epitaxial reactor and its use in the manufacture of standard thickness N and P wafers. Besides the advantage of having significantly reduced cost, these Epi wafers have high quality, better mechanical strength and resistance to light inducted degradation due to significantly reduced oxygen content.

INTRODUCTION

Crystalline silicon solar cells and modules continue to be the dominant photovoltaic technology, with a market share of more than 80%. The Si module prices have dramatically declined and are approaching below $0.60 /W for crystalline Si modules [1]. In order to further reduce the price, advances have to be made to enable higher module efficiencies and lower manufacturing costs. Significant progress has been reported in cell manufacturing technologies to get efficiencies of > 20%, for e.g. in the emitter formation[2], surface passivation[3], metallization[4], cell architecture including back contact solar cell[5], heterojunction solar cells[6,7] and thin silicon solar cells[8]. However, so far there has not been a production proven approach that reduces the cost while maintaining or increasing the efficiency.

Crystal Solar has pioneered the Direct Gas to Wafer[TM] technology to produce epitaxial kerfless crystalline silicon wafers. This technology bypasses the traditional steps of polysilicon production, Czochralski or Multicrystalline ingot growth and traditional wire-saw wafering. With a pilot manufacturing line, we have demonstrated this Direct Gas to wafer approach which essentially involves creating a release layer on a re-useable Si substrate, growing the wafer epitaxially, and mechanically separating the epitaxial wafer. The substrate is then cleaned and re-used multiple times so that its cost is insignificant. Figure 1 shows a schematic of this approach. Due to the kerfless nature of this process, the cost of these wafers can be up to 50%

lower than standard wafers. In addition, as will be shown, there are advantages with respect to lower oxygen concentration, increased uniformity in resistivity and better mechanical strength as compared with Cz wafers.

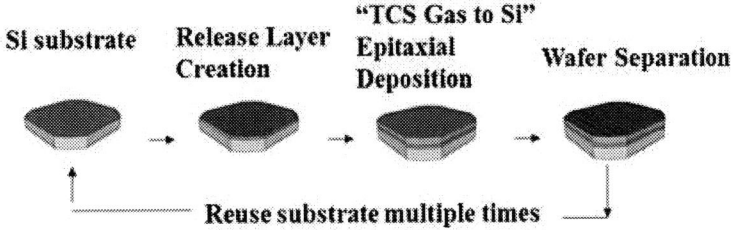

Figure 1 Schematic of the Crystal Solar's approach for epitaxial Si wafers

EXPERIMENT

The process starts with porous silicon release layer formation in chemically cleaned heavily doped p-type Si substrates with resistivity of ~0.01 ohm-cm. The porous Si layers are critical for high quality epitaxial growth and the release of the epitaxial Si wafer from the substrate. The key technical features associated with the porous Si release layer we have developed include multiple wafer processing, tailored multi-layer porosity, and uniformity of thickness and porosity across the wafer. Figure 2 shows Crystal Solar's porous silicon equipment. This pilot system can deliver > 200 wafers/hour scalable to > 1200 wafers/hour. Figure 3 shows a SEM cross section view of a 3-layer porous Si formed using this system. The low porosity creates a suitable template for high quality epitaxial growth, whereas the high porosity layer consolidates into voids which enable eventual release of the epitaxial wafer from the substrate. Crystal Solar has carefully engineered the porosity and the thickness of these layers to minimize substrate consumption while maximizing the yield of the epitaxial wafer separation from the substrate.

Figure 2 Crystal Solar's porous silicon equipment for multiwafer processing.

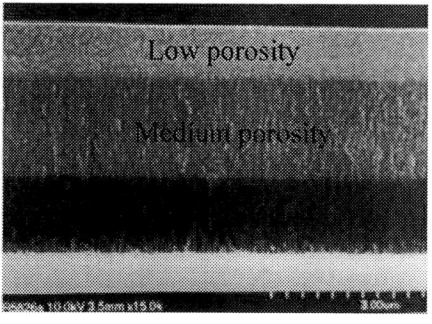

Figure 3 SEM cross section view of 3-layer porous Si formed using the pilot system in Figure 2.

The substrates with the porous Si are then loaded the high throughput silicon epitaxial reactor. Crystal Solar's patented epitaxial reactor has been designed to take advantage of non-linear depletion of Trichlorosilane (TCS) across the surface of the substrate to get very high growth rates of > 4um/min in a mini-batch (currently 24) of substrates. Running in the depletion mode also ensures a high conversion efficiency of TCS to Silicon (currently higher than 40%). To enable good thickness uniformity of the Epi Si wafers, in spite of the non linear depletion, the gas flow and temperature across the wafers have been optimized. Figure 4 show a picture of Crystal Solar's silicon epitaxial reactor with two chambers. These chambers feature a low volume due to the unique hardware design and this allows for multiple chambers to be supported by a single central robot. With this approach, throughputs > 300 wafers/hour can be obtained for thick Epi wafers.

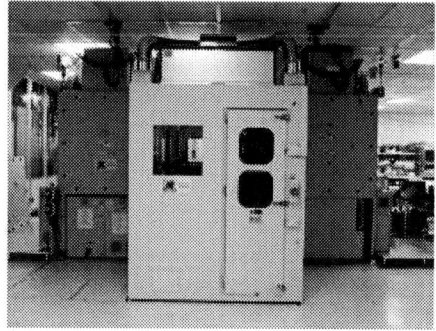

Figure 4 Crystal Solar's patented silicon epitaxial reactor with two chambers

To release the Epi wafer from the substrate, we have developed a laser based scribing and mechanical exfoliation system that automatically extracts the Epi wafer from the substrate. Figure 5 shows the laser system and the exfoliation process. Once the Epi wafer is released, the substrate is reclaimed by a simple chemical etch. We have demonstrated substrate reuse up to 50 times using a manual approach with no degradation in the lifetime of the Epi wafers and physical quality of the substrates.

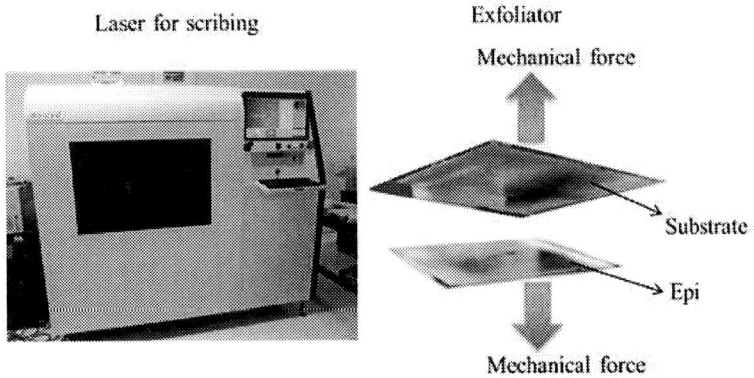

Figure 5 Crystal Solar's exfoliation system-laser and concept for Epi wafer release

DISCUSSION

Using the process described above, we have been generating both P-Type and N-Type Epi wafers with sizes of 156 mm x 156 mm and 125 mm x 125 mm full squares. Figure 6 shows these wafers with thickness of around 200 μm. Based on the characterization results, we demonstrate the considerable advantages that the Epi wafers have over conventional Cz solar wafers. These include well-controlled and repeatable doping concentration, no light induced degradation in cell, and higher mechanical strength. The results will be discussed in the following sections.

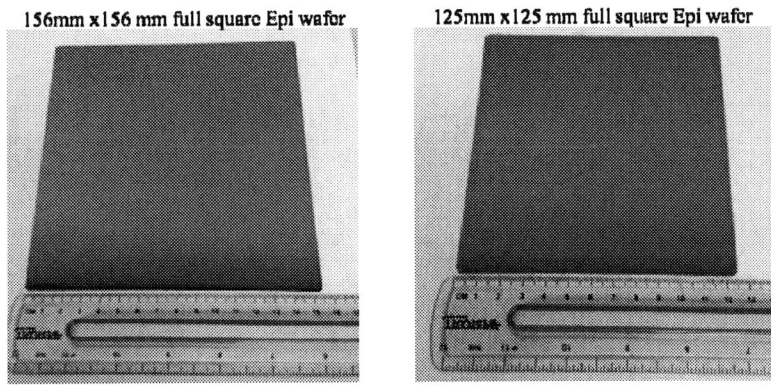

Figure 6 Pictures for 156 mm x 156 mm (Left) and 125 mm x 125 mm (Right) full square Epi wafers

Thickness uniformity

As discussed in the previous section, good thickness uniformity has been achieved in these epitaxial with the control of the gas flow and temperature. Figure 7 shows the thickness distribution within the 24 wafers grown in the same run. The average thickness is 69 μm with a 1 sigma deviation of 5.3% in the non-uniformity. The average growth rate is 4.8μm/min. We have maintained the thickness uniformity and growth rate for both P-Type and N-Type Epi wafers. The hardware design and growth approach allow for high growth rate and TCS utilization for various thicknesses ranging from 20 microns to 200 microns. Besides the thickness uniformity, we also have well-controlled and repeatable resistivity for both P-Type and N-Type Epi wafers. Figure 8 shows the resistivity and thickness uniformity of a 156 mm x 156 mm P-Type Epi wafer with standard wafer thickness. The epitaxial growth process also has the flexibility for different doping concentrations. The wide range of the growth thickness and doping concentration with the epitaxy allows for the implementation of different device architectures.

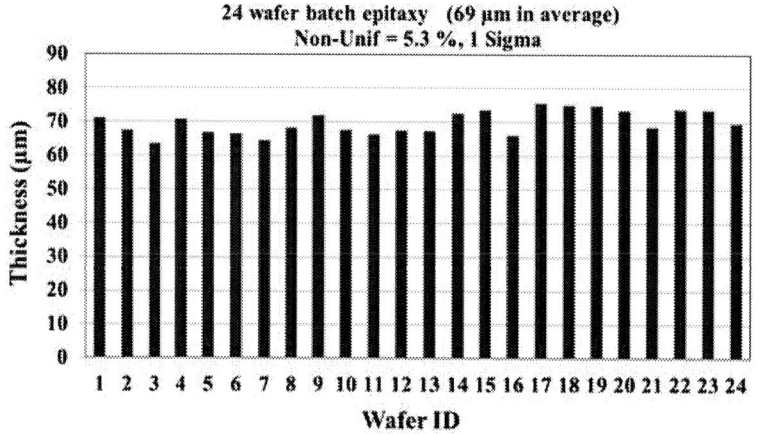

Figure 7 Thickness of a 24 wafer batch.

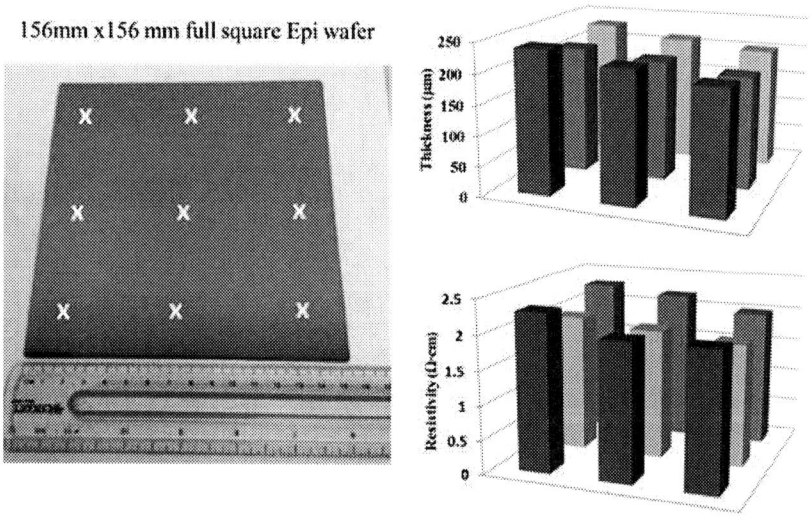

Figure 8 Thickness and resistivity of a 156 mm x 156 mm P-Type Epi wafer.

Effective lifetime of Epi wafers

We have investigated the effective minority carrier lifetime of the boron doped P-Type and phosphorus doped N-Type Epi wafers with different surface passivation techniques. The as-grown Epi wafers were separated from the substrate and chemically cleaned to remove the remaining porous Si. Figure 9 shows the representative microwave photoconductive decay (μ-PCD) lifetime maps measured with Semilab WT-2000PV for our P-Type and N-Type Epi wafers with a size of approximately 156 mm x 156 mm. The samples were passivated by an Iodine/Ethanol solution. Both maps suggest a tight distribution of the effective lifetime in each type wafer.

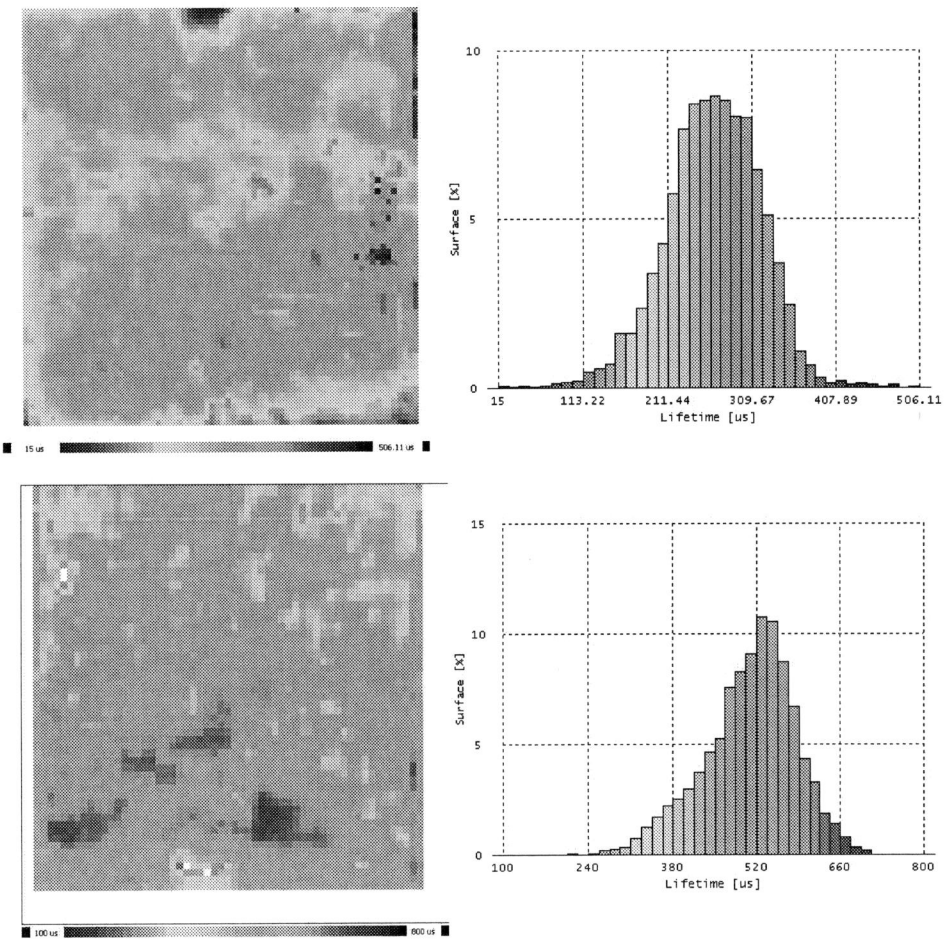

Figure 9 μ-PCD lifetime maps for Epi wafer with a size of approximately 156 mm x 156 mm: P-Type EPi wafer (Top) with a thickness of 200μm and resistivity of 2 Ω·cm, the red spot at the wafer edge can be due to the tweezer mark from handling; N-Type Epi wafer (Bottom) with a thickness of 120μm and resistivity of 1 Ω·cm.

We also measured the injection-dependent effective lifetime with a Sinton lifetime tester for our P-Type and N-Type Epi wafers. Since the minority carrier effective lifetime is strongly dependent on the surface condition, we performed a thorough cleaning and deposited different passivation materials on their surfaces. The P-Type Epi was 180 μm thick and had a resistivity of 2 Ω·cm, and was passivated with 20 nm of Al₂O₃ deposited on both sides. Effective minority carrier lifetime of >700 μs at $\Delta n = 10^{15}$ cm⁻³ has been achieved. The N-Type Epi sample was 120 μm thick and has a resistivity of 2 Ω·cm, passivated with amorphous silicon. Effective minority carrier lifetime >2 ms at $\Delta n = 10^{15}$ cm⁻³ has been achieved. It is clear that the minority carrier lifetimes we have achieved for our both P-Type and N-Type Epi wafers are well suited to achieve efficiencies above 21%. We also analyzed the dislocation density after a defect etch. The results have shown that the average dislocation density of the epi wafers is < 1E3/cm² which is insignificant in terms of its effect on the device performance.

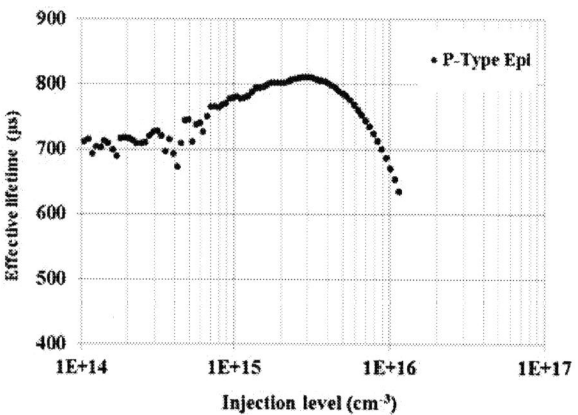

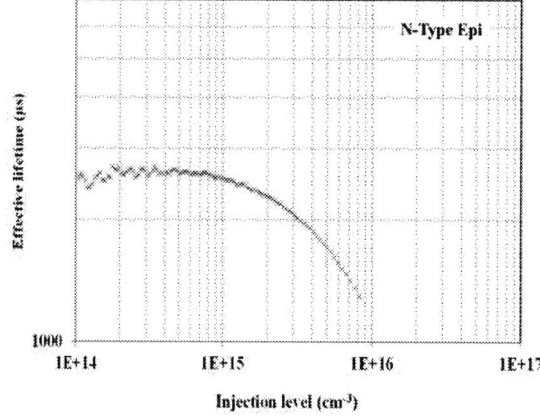

Figure 10 Injection-dependent effective minority-carrier lifetimes of P-Type (Top) and N-Type Epi wafers (Bottom)

Oxygen concentration in the Epi wafer

It is well known that conventional Cz grown Si wafers have high interstitial oxygen concentration in the range of 1E18 cm⁻³. High oxygen concentration leads to the formation of oxygen related defects, e.g. boron-oxygen complexes which leads to light induced degradation on P-type cells (LID), thermal donors and oxygen related precipitates [9]. Our Epi wafers have shown significantly lower oxygen concentration compared to the conventional Cz solar wafers. Figure 11 shows the SIMS measurement of the oxygen concentration in the boron doped P-Type Epi wafer. The sample was separated from the substrate and porous silicon residue was not etched off the Epi wafer. The measurement was performed from the porous silicon side of the Epi. The higher oxygen concentration noticed in the top surface layer may be due to the native oxide surface of the porous structure. However, as can be seen, the bulk P-Type Epi wafer indicates a lower oxygen concentration of approximately 3×10^{17} cm^{-3}.

In order to compare the LID of solar cells made with Cz solar wafers and our Epi wafers, we measured the efficiency degradation of three conventional P-Type Cz solar cells and three Epi solar cells made with our P-Type Epi wafers using the same standard fabrication process. The Epi solar cells have a structure of ARC/N-type emitter formed by POCl$_3$ diffusion in 90 μm thick boron doped p-Type Epi/p+ silicon substrate with screen printed contacts. All the solar cells have an area of 156 cm^2. The efficiencies were measured for as- processed solar cells and after 48 hours of one-sun light exposure. Figure 12 shows the relative efficiency degradation of the solar cells. The Epi cells show essentially 0% LID compared to ~3.5% relative efficiency degradation in the Cz cells. This lack of LID is directly attributed to the lower oxygen concentration in the Epi wafers.

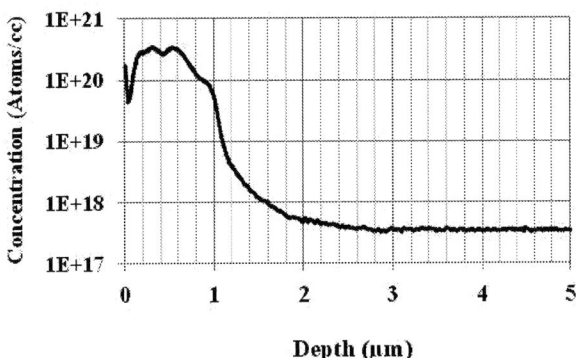

Depth (μm)

Figure 11 SIMS measurement of O for as grown P-Type Epi wafer

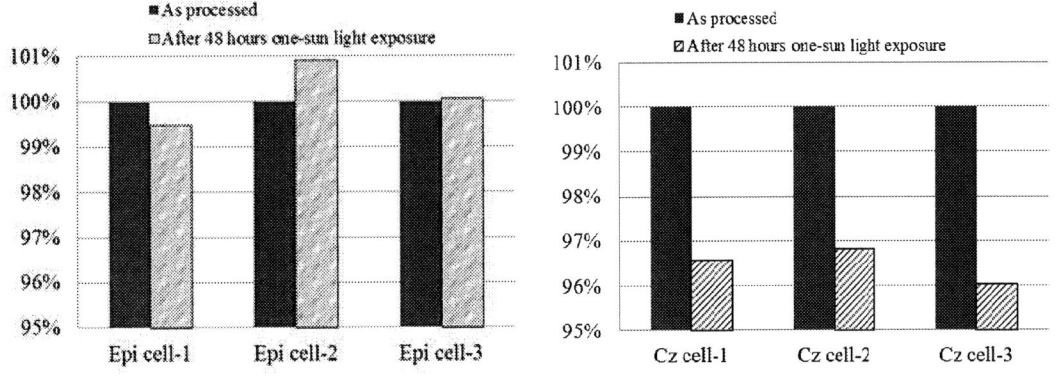

Figure 12 Relative efficiency of 3 Epi cell (Left) and 3 Cz solar cells (Right)
measured before and after 48 hours one-sun light exposure.

Mechanical strength

Epi wafers of various thickness ranging from 120 μm- 180 μm were tested for breaking strength using a standard Instron bend test. Figure 13 shows the break strength of Epi wafers versus standard CZ wafers and as can be seen the Epi wafers are 2-3x stronger. We believe that this results from the lack of any saw damage on the Epi wafers. Also these results imply that the Epi wafers can have a better chance of surviving the cell and module processing as the wafer thickness reduces for future generations of solar cells.

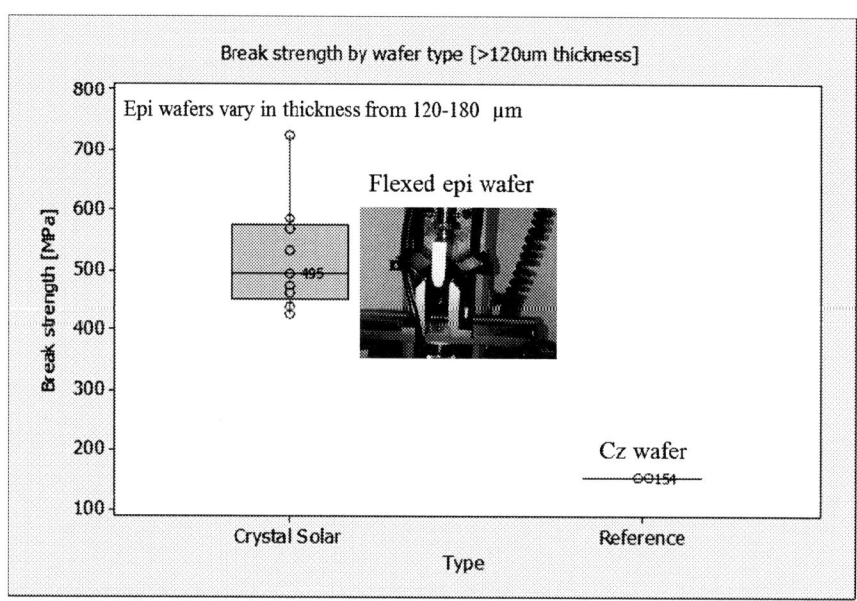

Figure 13 Break strength of Epi wafers versus standard Cz wafers

Solar cell efficiencies with Epi wafers

Both thick and thin silicon solar cells have been fabricated with our P-Type and N-Type Epi wafers. Figure 14 shows the current-voltage (I-V) curve of an Epi solar cell made on 180 μm P-Type Epi wafer with screen printed Ag front contact and full Al BSF and back contact. As can be seen, this efficiency is comparable to that of a standard Cz p-type cell. Furthermore, we have demonstrated an average efficiency > 20% for solar cells made with our N-Type Epi wafers. These N-type Epi wafers have a full square size and a lifetime of approximately 2 ms. The internal quantum efficiencies (IQE) of the N-Type Epi cell and N-Type Cz cell processed together match with each other very closely. The solar cell results confirm again the high bulk quality of both P-Type and N-Type Epi wafers. In addition to the high quality, the Epi wafers, due to their smooth surfaces, do not need a wet chemical etching to remove wire sawing damage which is required for conventional Cz solar wafers. Another obvious benefit of the Epi wafers is that they are square in shape compared to pseudo square Cz wafers, which contributes to the total area by ~ 2% and therefore increases the module power output by the same.

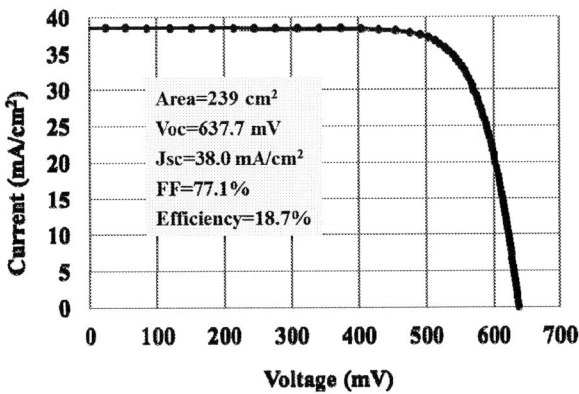

Figure 14 Current-voltage curve of an Epi solar cell made with 200 μm P-Type EPi wafer

Cost of Epi wafers

Given the kerfless nature of the Epi process, we expect the cost of these wafers to be about 50% lower than standard Cz wafers. The cost advantage and other benefits as discussed above of Epi wafers have attracted significant attention among PV cell manufacturers around the world and we expect rapid acceptance of these wafers into the PV market.

CONCLUSIONS

We have demonstrated the key features of a high throughput production silicon epitaxial reactor designed by Crystal Solar. With this reactor and the process flow developed for porous silicon release layer formation and exfoliation of Epi layer from the substrate, we are able to produce high quality full square N-Type and P-Type Epi wafers. These full square mono crystalline Epi wafers have well-controlled and repeatable doping concentration, high lifetime, better mechanical strength, and no light induced degradation in cell efficiency due to lower oxygen incorporation compared to conventional p-type wafers. The efficiencies for the P and N type cells made using Epi wafers are shown to be equivalent to that of the Cz wafers. We expect that the cost advantage of these wafers will enable rapid acceptance among the cell manufacturers.

ACKNOWLEDGMENTS

The authors would like to thank Douglas M. Powell from Massachusetts Institute of Technology for surface passivation and lifetime measurement; Chia-Wei Chen and Prof. Ajeet Rohatgi from Georgia Institute of Technology for P-Type solar cell making and measurement.

REFERENCES

1. http://www.greentechmedia.com/articles/read/Module-Costs-Dip-Below-50-Cents-Per-Watt-in-JinkoSolars-Strong-Q4

2. A. Rohatgi, D. L. Meier, B. McPherson, Y. Ok, A. D. Upadhyaya, J. Lai and F. Zimbardi. Energy Procedia **15**, 10-19, 2012.

3. G. Dingemans, M. C. M. van de Sanden, and W. M. M. Kessels. Electrochemical and Solid-State Letters, **13** (3) H76-H79, 2010.

4. L. Tous, R. Russell, J. Beckers, J. Bertens, E. Cornagliotti, P. Choulat, J.John, F. Duerinckx, J. Szlufcik, J. Poortmans, and R. Mertens. Proc. 28th EU PVSEC, 1008-1012, 2013.

5. P. J. Cousins, D. D. Smith, H. Luan, J. Manning, T. D. Dennis, A. Waldhauer, K. E. Wilson, G. Harley, and W. P. Mulligan. Proc. 35th IEEE PVSC, 000275 – 000278, 2010.

6. A. Yano, S. Tohoda, K. Matsuyama, Y. Nakamura, T. Nishiwaki, K. Fujita, M. Taguchi, and E. Maruyama. Proc. 28th EU PVSEC, 748-751, 2013.

7. E. Kobayashi, N. Nakamura, K. Hashimoto and Y. Watabe. Proc. 28th EU PVSEC, 691-694, 2013.

8. P. Kapur, M. M. Moslehi, A. Deshpande, V. Rana, J. Kramer, S. Seutter, H. Deshazer, S. Coutant, A. Calcaterra, S. Kommera, Y. Su, D. Grupp, S. Tamilmani, D. Dutton, T. Stalcup, T. Du, M. Wingert. Proc. 28th EU PVSEC, 2228-2231, 2013.

9. J. Horzel, L. Tous, A. Uruena De Castro, A. Seidl, R. Russell, E. Cornagliotti. Proc. 27th EU PVSEC, 780-788, 2012.

Mater. Res. Soc. Symp. Proc. Vol. 1666 © 2014 Materials Research Society
DOI: 10.1557/opl.2014.717

Optimization of the protocrystalline *p*-layer in *a*-Si:H-based *n-i-p* photodiodes

Y. Vygranenko[1,2], M. Fernandes[1,2], M. Vieira[1,2], A. Sazonov[3]

[1]Electronics, Telecommunications and Computer Engineering, ISEL, Lisbon, 1950-062, Portugal

[2]CTS-UNINOVA, Quinta da Torre, 2829-516, Caparica, Portugal

[3]Electrical and Computer Engineering, University of Waterloo, Waterloo, N2L 3G1, Canada

ABSTRACT

This work reports a carbon-free, blue-enhanced *a*-Si:H *n-i-p* photodiode with an optimized protocrystalline *p*-layer. Although the used deposition conditions for the *p*-layer correspond to the microcrystalline regime, thin layers are mostly protocrystalline due to the amorphous underlying undoped layer. This conclusion is supported by Raman spectroscopy measurements. We have also found that the optical band gap of the *p*-layer can be varied by adjusting the rf power. By widening the band gap and tuning the impurity concentration in the *p*-layer, absorption and recombination losses at the *p-i* interface were reduced. The current-voltage, capacitance-voltage, and spectral-response characteristics of fabricated photodiodes are correlated with the doping level, optical band gap, and deposition conditions for *p*-layers. The optimized device exhibits a leakage current of about ~80 pA/cm^2 at 5 V reverse bias. The external quantum efficiency reaches a peak value of 92% at a wavelength of 510 nm, and, at shorter wavelengths, decreases down to 66%@400nm.

INTRODUCTION

Hydrogenated amorphous silicon (*a*-Si:H) *p-i-n* photodiodes are commonly used as pixel sensors in digital radiographic flat-panel imaging detectors [1]. Photodiode performance is one of the factors limiting the signal-to-noise ratio and image quality. In particular, a high sensor sensitivity in the visible spectral range is required to provide an efficient optical coupling with conventional phosphors such as CsI:Tl or Gd$_2$O$_2$S:Tb [2]. One of the approaches to minimize the absorption losses in the *p*-layer is to use an *a*-Si$_{1-x}$C$_x$:H alloy having a wider band gap than *a*-Si:H [3-5]. However, the use of this technology in industry is limited because the most of production lines are for *a*-Si TFT backplanes, and the cost of their upgrade for additional doping gases may be unacceptably high.

In this work, we report on a carbon-free, blue-enhanced *n-i-p* photodiode by incorporating *p*-type doped protocrystalline silicon (pc-Si:H). In earlier studies, this material was applied to *a*-Si:H *n-i-p* solar cells to improve their performance through an increase in the open circuit voltage and a reduction in the series resistance [6-8]. Specifically, the thin (~20 nm) *p*-layers were prepared at low temperatures (< 200 °C) by PECVD using high hydrogen-to-silane flow ratios (typically R = [H$_2$] / [SiH$_4$] ~ 50 - 200) without crossing the thickness-dependent transition into the mixed-phase (amorphous + microcrystalline) growth regime. The growth of such *p*-layers follows a thickness evolution in which pure protocrystalline material is observed at the *p-i* interface and a low density of nanocrystallites nucleates with increasing thickness [9]. The microstructural evolution depends not only on the deposition conditions, but also on the micro-structure of the underlying *i*-layer needing the *p*-layer optimization in the actual cell configuration. In this work, the *p*-layer was optimized analyzing its impact on the device performance.

EXPERIMENT

Film and Device Deposition

The films and devices were deposited at 150°C onto Corning 1737 glass substrates using a multichamber, 13.56 MHz PECVD system, manufactured by MVSystems Inc. The distance between the substrate holder and the rf electrode, and the electrode area are 2.1 cm and 232 cm^2, respectively. Diborane (B_2H_6) and phosphine (PH_3), diluted in hydrogen to a concentration of 1%, were used as the doping gases.

The n-i-p photodiodes were fabricated by the following deposition sequence. First, a 200 nm thick Mo film was sputtered on the glass substrate, followed by the deposition of the n-i-p stack. Finally, a 65 nm thick ZnO:Al film was sputtered and patterned to form the top electrodes with areas ranging from 2×2 to 5×5 mm^2. The sheet resistance of the ZnO:Al layer is about 150 Ω/sq.

In order to avoid cross-contamination, the doped and undoped layers of the n-i-p stack were deposited in different chambers of the cluster tool system without breaking the vacuum. A 25 nm thick n-layer was prepared using a 1:4:0.01 mixture of SiH_4 / H_2 / PH_3. The deposition pressure and rf power were 500 mTorr and 2 W, respectively. Then, a ~500 nm of undoped a-Si:H layer was deposited in hydrogen-diluted silane plasma at $[H_2]$ / $[SiH_4]$ = 3, a deposition pressure of 400 mTorr, and an rf power of 2 W. Different samples were produced with p-layers grown in different growth regimes by varying an rf power from 2 to 90 W. The deposition pressure was kept at 1 Torr. The gas flow ratios, deposition rates, and layer thicknesses are shown in Table 1.

Table 1. Deposition conditions and thickness of the p-layer in the n-i-p stack.

Sample N°	rf power, W	$[H_2]$ / $[SiH_4]$	$[B_2H_6]$ / $[SiH_4]$	Dep. Rate, nm/s	Thickness, nm
#701	2	100	0.003	0.016	24
#702	10	100	0.005	0.044	22
#703	90	100	0.005	0.14	20

To study the microstructure of p-type material by Raman spectroscopy, bilayer test samples comprising a 30-nm-thick a-S:H buffer layer and a p-layer of various thicknesses were deposited on glass substrates. The buffer layer was deposited under the same deposition conditions that were used for the i-layer to reproduce the growth regime of the p-layer in the actual device configuration.

Characterization Techniques

Raman spectra were measured in the backscattering geometry using a Renishaw micro-Raman spectrometer with a 488 nm excitation laser line. A Dektak 8 surface profiler was used for film thickness measurements. Current-voltage characteristics of the photodiodes were measured at room temperature using a Keithley 4200-SCS semiconductor characterization system. The capacitance-voltage (C-V) characteristics of selected devices were measured at 1 kHz frequency using an Agilent 4284A LCR-meter. The spectral response measurements were performed with a PC-controlled setup based on an Oriel 77200 grating monochromator, a Stanford Research System SR540 light chopper, and an SR530 DSP lock-in amplifier. The system was calibrated in the spectral range of 300–1100 nm using a Newport 818-UV detector.

RESULTS AND DISCUSSION

Figure 1 shows the typical Raman spectra of glass/a-Si:H buffer/p-layer structures with p-type material deposited at an rf power of 90 W. Raman spectrum of the sample with a 30-nm-thick p-layer shows a broad band centered at 480 cm^{-1}, which is associated with the amorphous phase. A reference spectrum of a 150-nm-thick p-layer contains an additional narrow-band component centered at 516 cm^{-1}, originating from the nanocrystalline phase in the film [10]. Thus, the thin (< 30 nm) p-layer grown under the reported deposition conditions is protocrystalline in nature.

Figure 2 shows the typical quasi-static current-voltage characteristics of the n-i-p photodiodes. The forward and reverse bias sweeps were performed starting at zero bias. In order to minimize the transient current induced by the trapped charge in the i-layer, the sweep delay was set to 20 s, and the bias voltage was varied at 25 mV increments.

All photodiodes show an exponential dependence of the forward current over five orders of magnitude in the biasing range from 0.2 to 0.6 V. At higher biases, the series resistance and the space-charge limited current effect are factors defining the current-voltage dependence. The saturation current density (J_0) and diode ideality factor (n) values, determined through a fitting procedure, are shown in Table 2. The difference in the deduced n and J_0 can be explained considering the current components related to recombination in the i-layer bulk and at the p-i interface region [11]. Value $n = 1.55$ is associated with recombination dominated by the bulk, while the reduced $n \approx 1.3$ implies that the current is dominated by p-i interface recombination. Similar changes in the forward J-V characteristics of a-Si:H n-i-p cells attained by tailoring of the p-i interface have been reported elsewhere [12].

The fabricated photodiodes show low dark-state reverse bias leakage currents ranging from 40 to 80 pA/cm^2 at 5 V reverse bias. Note that the neighbor cells having different active areas (4 and 25 mm^2) exhibit about the same reverse current densities, i.e. the thermally generated bulk current and junction leakage current components are larger than the edge leakage current [1]. The magnitude of the leakage current is comparable to that in the state-of-the-art a-Si:H n-i-p photodiodes reported elsewhere [13].

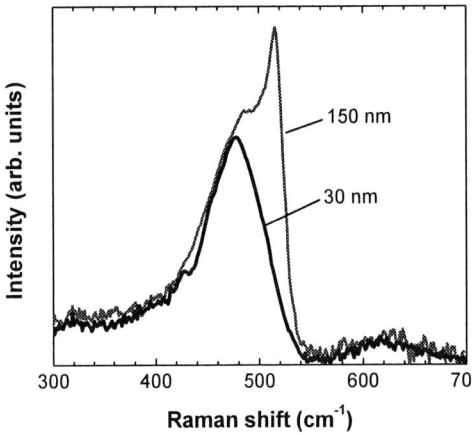
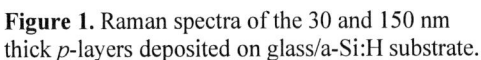

Figure 1. Raman spectra of the 30 and 150 nm thick p-layers deposited on glass/a-Si:H substrate.

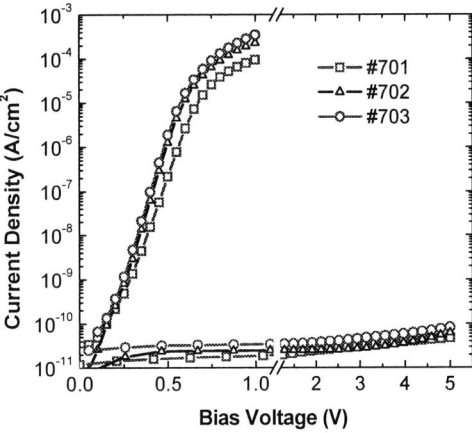

Figure 2. Current-voltage characteristics of n-i-p photodiodes.

Table 2. The diode ideality factor (n), saturation current density (J_0), depletion width at zero bias (W_0), and acceptor density deduced from current-voltage and capacitance-voltage characteristics.

Sample N°	rf power, W	n	J_0, fA/cm^2	W_0, nm	N_A, ×10^{18} cm^{-3}	α_{400nm}, ×10^5 cm^{-1}
#701	2	1.55	600	550	0.75	5.0
#702	10	1.30	330	516	1.44	2.9
#703	90	1.31	550	480	1.42	1.3

For more detailed study of the *i-p* interface, the capacitance-voltage characteristics were measured. Under moderate reverse bias the *i*-layer is fully depleted; therefore, the small decrease in the diode capacitance with increasing reverse voltage is due to expansion of the depleted region into the doped layers. In the devices under study, the conductivity of phosphine-doped *a*-Si:H, $\sigma \approx 10^{-3}$ S/cm, is much higher than that of *p*-type material, leading to a an asymmetrical extension of the depletion region, mainly in to the p-layer . In this case, the increase in the depletion width $\Delta W_P = W(V) - W_0$ within *p*-layer is

$$\Delta W_P = \varepsilon_0 \varepsilon \ A \left(\frac{1}{C(V)} - \frac{1}{C_0} \right), \tag{1}$$

where ε is the static dielectric constant of the semiconductor material, ε_0 the permittivity of free space, and A the area of the junction. Figure 3 shows a $\Delta W_P - V$ plot for fabricated samples. The dependences are close to linear in some biasing range above 1 V, when the *i*-layer is fully depleted. Here, the curve slope depends on the doping level and net capacitance. The acceptor density N_A can be estimated using expression

$$N_A = \frac{C(V)}{e \cdot A} \left(\frac{\delta W_P}{\delta V} \right)^{-1}, \tag{2}$$

where e is the electron charge. The obtained N_A values are shown in Table 2.

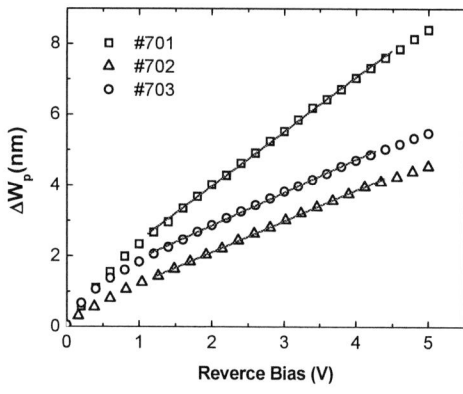

Figure 3. Variation of depletion width in the *p*-layer obtained from C-V measurements.

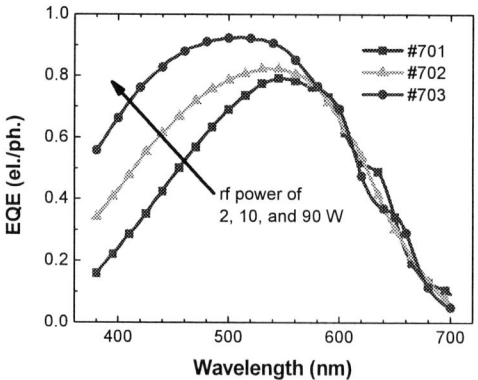

Figure 4. Spectral-response characteristics measured at a reverse bias of 5 V.

Figure 4 shows the external quantum efficiency (EQE) spectra of the photodiodes at a reverse bias of 5 V. Peak values of the curves vary from 79 to 92% in the narrow spectral interval of 510–545 nm. The photodiodes exhibit about the same spectral response in the long-wavelength region. Here, the observed mismatch of interference fringes is caused by a slight difference in the *n-i-p* stack thicknesses. At $\lambda < 580$ nm, the spectral response strongly depends on the growth regime of the *p*-layer. Apparently the density of Si nanocrystallites in pc-Si:H increases with an increasing rf-power leading to the observed absorption loss reduction. In particular, the absorption coefficient, α at 400nm, decreases from 5×10^5 to 1.3×10^5 cm^{-1} within the 2 to 90 W rf-power range (see Table 2). Note that the deduced values are lower than $\alpha_{400nm} = 6 \times 10^5$ cm^{-1} for device-quality a-Si:H, but they are similar to that for B-doped nc-Si:H films reported elsewhere [14]. Such materials with small (few nanometers in size) Si nanocrystallites exhibit optical gap widening due to quantum size effect.

Besides the light absorption in the *p*-layer, the *p-i* interface recombination can be also considered as a limiting factor for decreasing short-wavelength response. To analyze the recombination losses, the quantum efficiency measurements were performed at various biasing conditions. Figure 5 shows the EQE(−5V) / EQE(0V) ratios as a function of the wavelength. Also the EQE(V) / EQE(0V) curves measured at a wavelength of 400 nm are shown in Fig. 6. Sample #703 with the *p*-layer grown in the high-rf-power regime shows the lowest EQE enhancement at short wavelengths indicating reduced *p-i* interface recombination. At long-wavelengths, when the generation profile is quasi-uniform, the EQE ratio indicates the intensity of recombination within the *i*-layer bulk. The observed increase in the bulk recombination with increasing rf power points to a decrease in the built-in potential (V_{bi}) which depends on the bandgap alignment between the *i*- and *p*-layers. This conclusion is in good agreement with the open circuit voltages measured under AM1.5 illumination, decreasing from 0.82 to 0.68 V for samples with *p*-layers deposited at 2 to 90 W rf-powers, respectively. Besides, the $\Delta W_P - V$ dependence is strongly non-linear at low biases for Sample #703 (see Fig. 3), because the non-depleted region within the *i*-layer is wider due to the decreased V_{bi}.

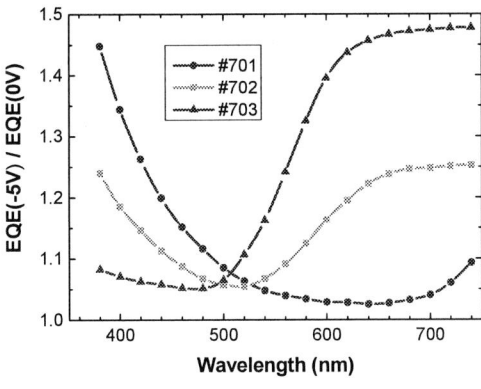

Figure 5. Ratio of EQE@-5V to EQE@0V as a function of wavelength.

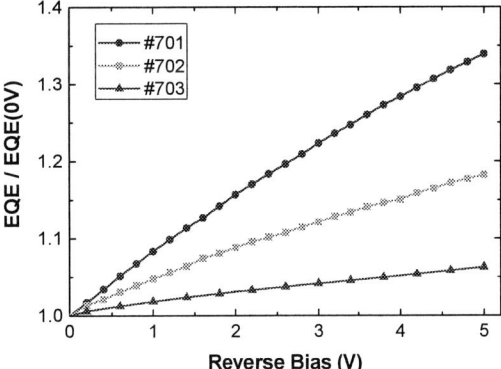

Figure 6. Variation of EQE with an increasing reverse bias at 400 nm wavelength.

CONCLUSIONS

Applying B-doped protocrystalline silicon grown in hydrogen-diluted plasma as a p$^+$ material, a series of *a*-Si:H-based *n-i-p* photodiodes has been fabricated and characterized. The proto-crystalline nature of thin *p*-layers is confirmed by Raman spectroscopy measurements. It is proven that the optical band gap of *p*-type pc-Si:H can be varied by adjusting the rf power. By widening the band gap and tuning the impurity concentration, the *p*-layer was optimized targeting both low-leakage current and high short-wavelength sensitivity. The optimized device (Sample #703) exhibits a leakage current of about 80 pA/cm^2 at a reverse bias of 5 V, and 66% EQE at a wavelength of 400 nm. The recombination losses in the *i*-layer bulk and at the *p-i* junction were estimated and correlated to the bandgap offset at the *pc*-Si:H/*a*-Si:H interface.

ACKNOWLEDGMENTS

The authors are grateful to the Portuguese Foundation of Science and Technology through research Project PTDC/EEA-ELC/115577/2009 for financial support of this research, and to the Giga-to-Nanoelectronics Centre at the University of Waterloo for providing some necessary equipment and technical help to carry out this work.

REFERENCES

[1] R. A. Street, Ed., *Technology and Applications of Amorphous Silicon* (Berlin: Springer-Verlag, 2000).
[2] J. Beutel, H. L. Kundel, and R. Van Metter, Eds., *Handbook of Medical Imaging*, (Washington, DC.: SPIE Press, 2000).
[3] H. Stiebig, F. Siebke, W. Beyer, C. Beneking, B. Rech, and H. Wagner, *Sol. Energy Mater. Sol. Cells* **48**, 351 (1997).
[4] P. Servati, Y. Vygranenko, A. Nathan, S. Morrison, and A. Madan, *J. Appl. Phys.* **96**, 7578 (2004).
[5] Y. Vygranenko, J. Chang, A. Nathan, *IEEE J. Quantum Electron.* **41**, 697 (2005).
[6] R. J. Koval, J. M. Pearce, Chi Chen, G. M. Ferreira, A. S. Ferlauto, R. W. Collins, and C. R. Wronski, Mater. Res. Soc. Symp. Proc. **715**, paper A6.1 (2002).
[7] R. J. Koval, Chi Chen, G. M. Ferreira, A. S. Ferlauto, J. M. Pearce, P. I. Rovira, C. R. Wronski, and R. W. Collins, *Appl. Phys. Lett.* **81**, 1258 (2002).
[8] G.M. Ferreira, Chi Chen, R.J. Koval, J.M. Pearce, C.R. Wronski, R.W. Collins, *J. Non-Crystal. Solids* **338–340**, 694 (2004).
[9] J. M. Pearce, N. Podraza, R. W. CollinsJ, M. M. Al-Jassim, K. M. Jones, J. Deng and C. R. Wronski, *J. Appl. Phys.* **101**, 114301 (2007).
[10] C. Smit, R.A.C.M.M. van Swaaij, H. Donker, A.M.H.N. Petit, W.M.M. Kessels and M.C.M. van de Sanden, *J. Appl. Phys.* **94**, 3582 (2003).
[11] J. Deng and C. R. Wronski, *J. Appl. Phys.* **98**, 024509 (2005).
[12] J. M. Pearce, R. J. Koval, A. S. Ferlauto, R. W. Collins, and C. R. Wronski, *Appl. Phys. Lett.* **77**, 3093 (2000).
[13] J. A. Theil, *Mat. Res. Soc. Symp. Proc.* **762**, paper A21.4 (2003).
[14] E. Fathi, Y. Vygranenko, M. Vieira, and A. Sazonov, *Appl. Surf. Science* **257**, 8901 (2011).

Mater. Res. Soc. Symp. Proc. Vol. 1666 © 2014 Materials Research Society
DOI: 10.1557/opl.2014.718

Near-UV background as a bridge between visible and infrared communication

M. Vieira[1,2,3], M. A. Vieira[1,2], V. Silva [1,2], I. Rodrigues[1], P. Louro[1,2]

[1]Electronics Telecommunication and Computer Dept. ISEL, R. Conselheiro Emídio Navarro, 1959-007 Lisboa, Portugal
[2] CTS-UNINOVA, Quinta da Torre, Monte da Caparica, 2829-516, Caparica, Portugal.
[3] DEE-FCT-UNL, Quinta da Torre, Monte da Caparica, 2829-516, Caparica, Portugal

ABSTRACT

In this paper we present a monolithically integrated wavelength selector based on a multilayer pi'n/pin a-SiC:H integrated optical filter that requires appropriate near-ultraviolet steady states optical switches to select the desired wavelengths in the VIS-NIR ranges.

Results show that the background intensity works as a selector in the infrared/visible regions, shifting the sensor sensitivity. Low intensities select the NIR range while high intensities select the visible part accordingly to its wavelength. Here, the optical gain is very high in the red range, decreases in the green range, and stays near one in the blue region decreasing strongly in the near-UV range. The transfer characteristics effects due to changes in steady state light intensity and wavelength backgrounds are presented. The relationship between the optical inputs and the output signal is established when a multiplexed signal is analyzed.

INTRODUCTION

Newly developed technologies for infrared telecommunication systems allowed the increase of capacity, distance and functionality, switching and control with the design of new reconfigurable logic active filter gates by "bridging the gaps" and combining the optical filters properties [1]. Expanding far beyond traditional applications in optical interconnects at telecommunication wavelengths [2, 3], the SiC nanophotonic integrated circuit platform has recently proven its merits for working with visible range optical signals. To enhance the transmission capacity and the application flexibility of optical communication efforts have to be considered, namely the Wavelength Division Multiplexing based on tandem a-SiC:H light controlled filters, when different visible signals are encoded in the same optical transmission path [4, 5]. In this paper, the shift of the visible range to telecom band can be accomplished using the same wavelength selector but under appropriated near ultraviolet optical bias, acting as reconfigurable active filters in the visible and near infrared ranges. These active filters act as interface devices that establish the bridge between the infrared and red spectral range playing a key role to bridging the infrared and the visible optical communication technology. They can be used to realize different filtering processes such as: amplification, switching, and wavelength conversion.

DEVICE DESIGN, CHARACTERIZATION AND OPERATION

The light tunable filter is realized by using a double pi'n/pin a-SiC:H photodetector with TCO front and back biased optical gating elements as depicted in Figure 1. The active device consists of a p-i'(a-SiC:H)-n/p-i(a-Si:H)-n heterostructure. The thicknesses and optical gap of the front i'-(200 nm; 2.1 eV) and back i- (1000 nm; 1.8 eV) layers are optimized for light absorption in the blue and red ranges, respectively [6]. Optoelectronic characterization was performed through spectral response and transmittance measurements without and with steady state applied optical

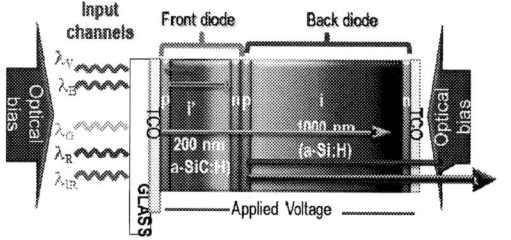

bias. The optical bias (ϕ; background) was superimposed using near-ultraviolet LEDs (350 nm-400 nm). Currents between 1mA and 30 mA were used to drive the LEDs in order to change the light flux background.

Monochromatic (infrared, red, green, blue and violet ; $\lambda_{IR,R,G,B,V;}$) pulsed communication channels (input channels) are combined together, each one with a specific bit sequence and absorbed accordingly their

Figure 1 Device configuration and operation.

wavelengths (see arrow magnitudes in Figure 1). The combined optical signal (multiplexed signal; MUX) is analyzed by reading out the generated photocurrent under negative applied voltage (-8V), without and with near ultraviolet background (350<λ<400nm) and different intensities, applied either from front (λ_F) or back (λ_B) sides. The device operates within the visible range using as input color channels the square wave modulated low power light supplied by near-infrared/red (NIR/ R: 880 nm-626nm), green (G: 524 nm), blue (B: 470 nm) and violet (V: 400 nm) LEDs. In Figure 2a the transmittances from the front and back diodes are plotted as well as the transmittance of the complete device without any background light. In Figure 2b the transmittance is displayed under different 390 nm background intensities.

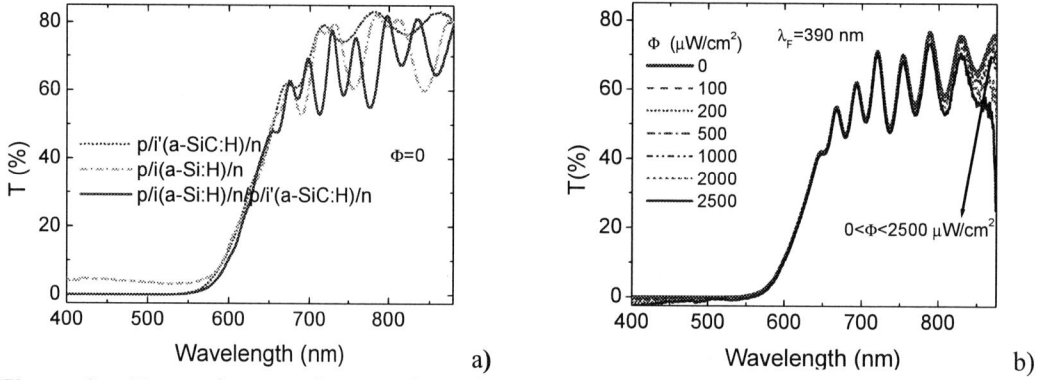

Figure 2 Transmittances from: a) front, back and whole device; b) the pi'npin structure under front irradiation with 390 nm irradiation and different intensities.

Results confirm the influence of the thickness of each front and back diode on the transmittance of the whole device. It is interesting to notice that under front light irradiation the transmittance decreases in the infrared range as the background intensity increases leading to an infrared absorption window.

SPECTRAL SENSITIVITY

The spectral sensitivity was tested through spectral response measurements [7] without and under 350 nm and 400nm front and back backgrounds of variable intensities. In Figure 3 the spectral gain (α), defined as the ratio between the spectral photocurrent with and without applied optical bias, is displayed under near-UV (λ=350 nm; Figures 3a and 3b) and violet (λ=400 nm;

Figures 3c and 3d) illuminations. In Figure 3a the light was applied from the front (λ_F) and in Figure 3b the irradiation occurs from the back side (λ_B) while in Figures 3c and Figure 3d visible violet light was used from front and back sides, respectively. The background intensity (ϕ) was changed between $5\mu Wcm^{-2}$ and $3800 \mu Wcm^{-2}$.

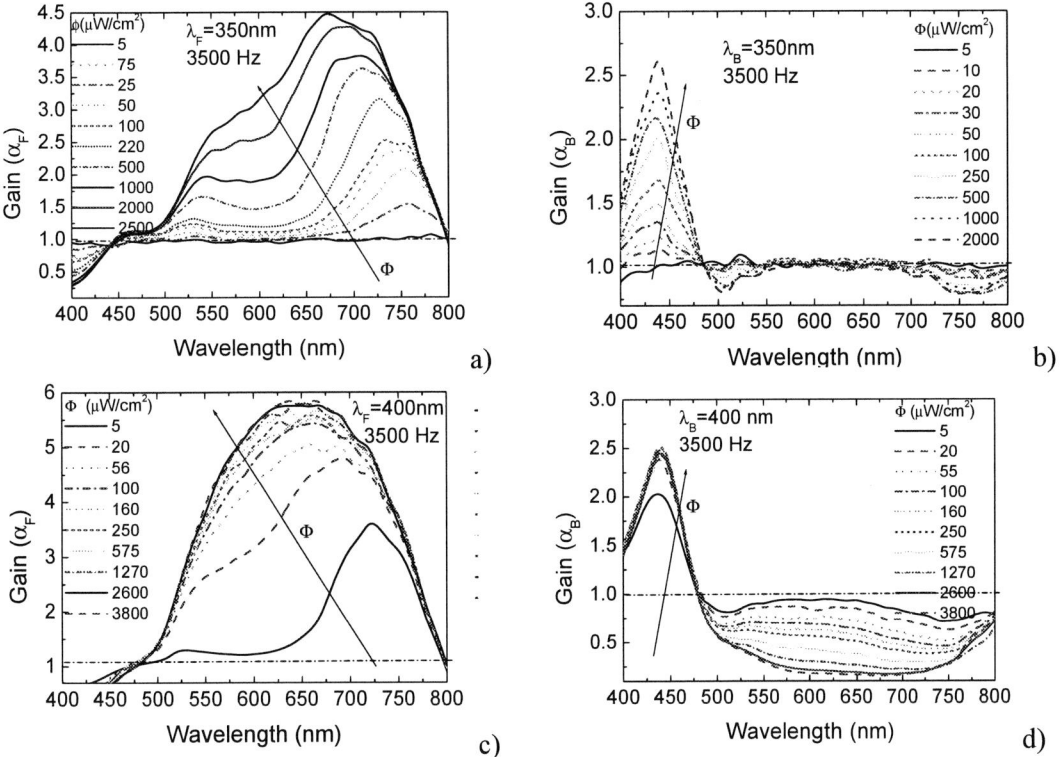

Figure 3 Front (λ_F, left) and back (λ_B, right) spectral gains ($\alpha_{F,B}$) under $\lambda=350$ nm (top) and $\lambda=400$ nm (bottom) irradiation.

Results show that the optical gains have opposite behaviors under front and back irradiations. Under 350 nm front irradiation (Figure 3a) and low flux, the gain is high in the infrared region, presents a well-defined peak at 750 nm and strongly quenches in the visible range. As the power intensity increases the peak shifts to the visible range and can be deconvoluted into two peaks, one in the red range that slightly increases with the power density of the background and another in the green range that strongly increases with the intensity of the UV radiation. In the blue range the gain is much lower. This shows the controlled high-pass filtering properties of the device under different background intensities. Under back bias (Figure 3b) the gain in the blue/violet range has a maximum near 420 nm that quickly increases with the intensity. Besides it strongly lowers for wavelengths higher than 450 nm, acting as a short-pass filter. Thus, back irradiation, tunes the violet/blue region of the visible spectrum whatever the flux intensity, while front irradiation, depending on the background intensity, selects the infrared or the visible spectral ranges. Here, low fluxes select the near infrared region and cuts the visible one, the reddish part of the spectrum is selected at medium fluxes, and high fluxes tune the red/green ranges with different gains. Under visible violet front and back backgrounds (Figures 3c and 3d) the behavior is similar, however under front irradiation the spectral sensitivity quickly increases and saturates at low fluxes when compared with the near-UV irradiation. Under back light the visible range is strongly quenched as the flux intensity increases.

VISIBLE AND INFRARED TUNING

Three monochromatic pulsed lights separately (645nm, 697 nm, 880 nm input channels,) or combined (MUX signal; Figure 4) illuminated the device at 12000 bps. Steady state 390 nm bias at different intensities ($5\mu Wcm^{-2}<\phi_{F,B}<3000\mu Wcm^{-2}$) were superimposed separately from the front and the back device side and the photocurrent was measured. The intensity to drive the LEDs was adjusted to generate almost the same intensities without applied optical bias. In Figure 4a, the transient signals are presented under front irradiations and in Figure 4b under back light. The current level was set to zero when all the input channels were off. On the top, the signals used to drive the input channels are shown to guide the eyes into the *on/off* channel states.

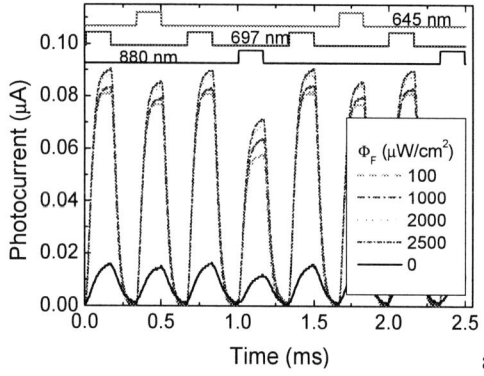 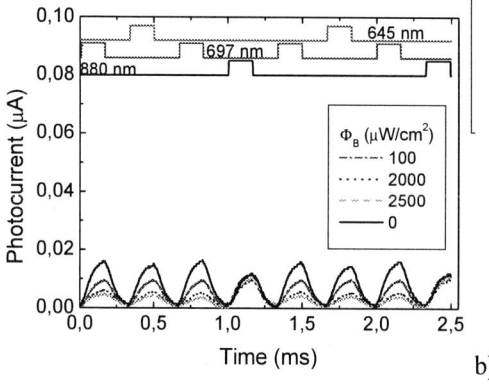

Figure 4 Front (a) and back (b) spectral photocurrent signal using λ=390 nm irradiation at different intensities.

As expected from Figure 3, in the red/infrared spectral ranges, the optical gain depends on optical bias intensity and on the wavelength of the input channels. Results show that, even under transient conditions and using commercial visible and NIR LEDs, the background side and intensity alters the signal magnitude of the input channels. Under front irradiation, as the light flux increases, the magnitudes of all the input channels increases being higher at 697 nm then at 645 nm or 880 nm. Under back irradiation, as the flux intensity increases the magnitude of the channels decreases, quickly in the visible range and stays almost constant in the infrared range. This nonlinearity provides the possibility for selective tuning of the visible and IR wavelengths allowing their recognition. In Table 1 the optical gains for the individual input channels are displayed under front (α_F) and back (α_B) irradiation. In Figure 5, the MUX signals due to the combination of the same wavelength channels but under two different bit sequences are displayed, under front and back irradiation. The signals were normalized to their values without background. In Figure 5a none of the channels are simultaneously *on* while in Figure 5b an overlap between the 880nm and the 697 nm channels occurs. Results show that by shining 390 nm light, separately, from both sides and using appropriate intensity, it is possible to tune different wavelengths in the VIS-NIR range.

Table 1 Optical gains under front and back irradiation.

Input channel (nm)	α_F	α_B
645	4	0.3
697	5	0.3
880	3.7	0.8

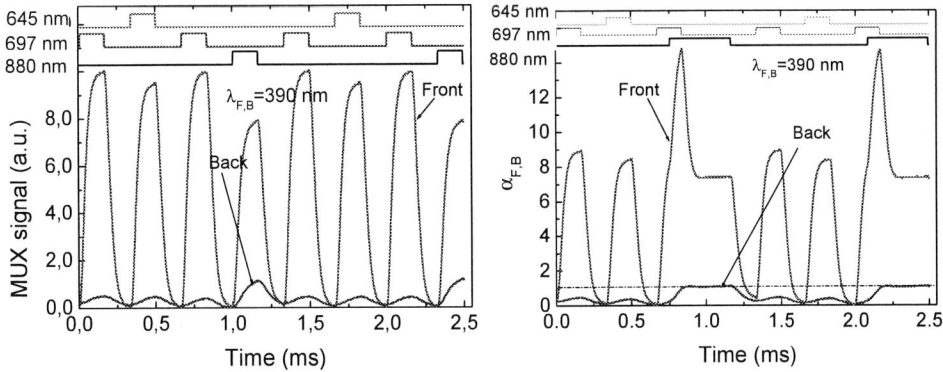

Figure 5   Front and back MUX signals under front and back λ=390 nm irradiation and different bit sequences.

Under front and back irradiation the gain of the three input channels is different ($\alpha_{F,B}$, Table 1). This nonlinearity allows identifying the different input channels in a narrow red/infrared range. Near-UV radiation is absorbed at the beginning of the front diode and, due to the self-bias effect, increases the electric field at the back diode where the red/infrared incoming photons (see Figure 2) are absorbed accordingly to their wavelengths (see Figure 3) resulting in an increased collection. Under back irradiation the electric field decreases mainly at the i-n back interface quenching the red input signals. The infrared signal stays near its value without back bias. This effect may be due to the increased absorption under back irradiation (Figure 2) that increases the number of carriers generated by the infra-red photons. So, by switching between front to back irradiation the photonic function is modified from a long- to a band-pass filter allowing, alternately selecting the red or the infrared channels, making the bridge between the visible and the infrared regions.

CODER/DECODER DEVICE

In Figure 6, the MUX signals due to the combination of the red, green, blue and violet visible input channels is displayed under front and back irradiations. On top the signals used to drive the input channels are also displayed. Figure 6a shows the normalized MUX signals due to the combination of all the sixteen possible input channels. In Figure 6b a random combination is presented to test the decoding methodology. Under front or back irradiation, each of those four channels, by turn, is enhanced or quenched (Figure 3). So, 2^4 ordered levels are detected (horizontal dotted lines) and grouped into two main classes due to the high amplification of the red channel ($\alpha_{626}=5.5$). The upper eight (2^3) levels are ascribed to the presence of the red channel (R=1), and the lower eight to its absence (R=0), allowing to decode the red channel. Since under front irradiation the green channel is also amplified ($\alpha_{524}=2.5$) the four (2^2) highest levels, in both classes, are ascribed to the presence of the green channel (G=1) and the four lower ones to its lack (G=0). The blue channel is slightly amplified ($\alpha_{470}=1.1$). So, in each group of 4 entries, two subclasses (2^1) can be found: the two set of higher levels correspond to the presence of the blue channel (B=1) and the two lowers to its absence (B=0). Finally, each group of 2 entries have two very near sublevels, the higher where the violet channels is ON (V=1) and the lower where it is missing (V=0). In the right side of the figure the sixteen RGBV sublevels are inserted. Under back irradiation, the violet channel is strongly enhanced, the blue channel slightly and the green and red reduced (Fig. 3b). So, from the front and back information the different bit sequences were decoded and the signal demultiplexed [7].

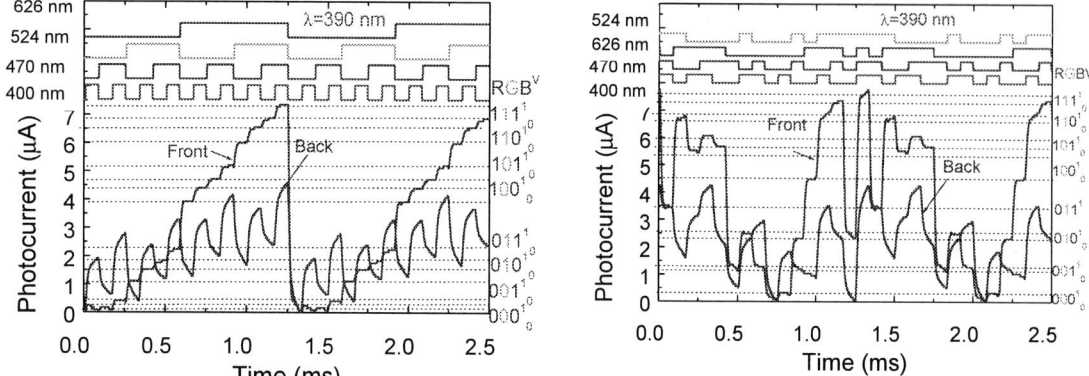

Figure 6. MUX/DEMUX signals under 390 nm front and back UV irradiation and decoded RGBV binary bit sequences. On top the signals used to drive the input channels are shown.

CONCLUSIONS

An optoelectronic device based on a-SiC:H technology is analyzed. Tailoring the filter wavelength in the NIR/VIS was achieved by using near-ultraviolet backgrounds and changing the irradiation side and intensity. Results show that the pi′n/pin multilayered structure becomes reconfigurable under front and back irradiation, acting as data selector in the VIS/NIR ranges. The device performs WDM optoelectronic logic functions providing photonic functions such as signal amplification, filtering and switching.

ACKNOWLEDGEMENTS

This work was supported by FCT (CTS multi annual funding) through the PIDDAC Program funds and PTDC/EEA-ELC/111854/2009 and PTDC/EEA-ELC/120539/2010.

REFERENCES

[1]   Z. Zhongwen, 2008 *Jornal of Multimedia Ubiquitous Engineering*, **3**, NO. 4 , pp.17-24.

[2]   P.P. Yupapin, P. Chunpang 2009, *Int. J. Light Electron. Opt.*, **120** (18) pp.976-979.

[3]   S. Ibrahim, L. W. Luo, S. S. Djordjevic, C. B. Poitras, I. Zhou, N. K. Fontaine, B. Guan, Z. Ding, K. Okamoto, M. Lipson, and S. J. B. Yoo, 21 Mar 2010, paper OWJ5. *Optical Fiber Communications Conference*, OSA/OFC/NFOEC, San Diego.

[4]   M. Vieira, P. Louro, M. Fernandes, M. A. Vieira, A. Fantoni and J. Costa Advances in Photodiodes, InTech, Chap.19, pp:403-425 (2011).

[5]   M. A. Vieira, M. Vieira, J. Costa, P. Louro, M. Fernandes, A. Fantoni, Sensors & Transducers Journal, 9, Special Issue, December 2010, pp.96-120.

[6]   M.A. Vieira, P. Louro, M. Vieira, A. Fantoni, A. Steiger-Garção, 2012 *IEEE sensor jornal*, **12**, NO. 6, pp. 1755-1762.

[7]   M. A. Vieira, M. Vieira, P. Louro, V. Silva and A. S. Garção, 2013 *Journal of Physics: Conference Series* **421** 012011 (March 2013) doi:10.1088/1742-6596/421/1/012011

Increased sensitivity in a-SiC pinpin multilayers in the VIS-NIR range under UV light

V. Silva[1,2], I. Rodrigues[1], M. A Vieira[1,2], P. Louro[1,2], M. Vieira[1,2,3]

[1]Electronics Telecommunication and Computer Dept. ISEL, R. Conselheiro Emídio Navarro, 1959-007 Lisboa, Portugal
[2] CTS-UNINOVA, Quinta da Torre, Monte da Caparica, 2829-516, Caparica, Portugal.
[3] DEE-FCT-UNL, Quinta da Torre, Monte da Caparica, 2829-516, Caparica, Portugal

ABSTRACT

In this paper we experimentally demonstrate the use of near-ultraviolet steady state illumination to increase the spectral sensitivity of a double a-SiC/Si pi'n/pin photodiode beyond the visible spectrum (400 nm-880 nm). The concept is extended to implement a one by four wavelength division multiplexer with channel separation in the visible/near infrared range.
Optoelectronic characterization of the device is presented and shows the feasibility of tailoring the wavelength and bandwidth of a polychromatic mixture. Several monochromatic pulsed lights in the VIS/NIR range, separately or in a polychromatic mixture illuminated the device. Independent tuning of the wavelengths is performed by steady state 390 nm optical bias superimposed from front and back sides. Results show that, front background enhances the light-to-dark sensitivity of the medium, long and infrared wavelength channels, and quench strongly the shorter wavelengths. Back background has the opposite effect; it only enhances the channel magnitude in short wavelength range and strongly reduces it in the long ones. This nonlinearity provides the possibility for selective tuning a specific wavelength. A capacitive optoelectronic model supports the experimental results. A numerical simulation is presented.

INTRODUCTION

Optical Wireless Communication, in the infrared and visible range, is an attractive solution, especially in environment settings where radio communication encounters difficulties. Visible Light Communication (VLC) has several advantages: Light communication is visible, so it is easy to determine who can listen or receive a message. A side effect is that light communication does not require part of the radio spectrum and can therefore be seen as a suitable extension in bandwidth-limited scenarios. Visible light is present in many places, so there is the opportunity to combine light communication with lighting design to let VLC coexist with, or even benefit from, the lighting setup present in many offices, homes, or institutions. The VLC principle is a relatively new approach for optical free space applications. However, it has been so far considered mainly for internet access or home networks, but more applications are feasible [1, 2, 3].

SiC active filter based on a-Si technology, has recently proven its merits to operate with visible optical signals. To enhance the transmission capacity and the application flexibility of the optical communication efforts have to be considered, namely the fundamentals of Wavelength Division Multiplexing based on a-SiC:H light controlled filters when different visible signals are encoded in the same optical transmission path [4, 5]. Those active filters can be used to perform different filtering processes such as: amplification, switching, and wavelength conversion. In this paper we demonstrate the use of near-UV steady state illumination to increase the spectral sensitivity of a double a-SiC/Si pi'n/pin photodiode beyond the visible spectrum.

DEVICE DESIGN, CHARACTERIZATION AND OPERATION

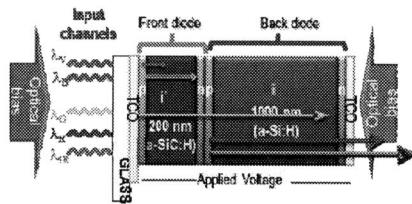

Figure 1 Device structure and operation.

The sensor is a two stacked p-i-n structures (p(a-SiC:H)- i'(a-SiC:H)-n(a-SiC:H)-p(a-SiC:H)-i(a-Si:H)-n(a-Si:H)) sandwiched between two transparent contacts one at each end. The thicknesses and optical gap of the i'- (200nm; 2.1 eV) and i- (1000nm; 1.8eV) layers are optimized for light absorption in the blue and red ranges [5], respectively. Based on silicon carbon technology this structure can be seen in Figure 1, where λ_V, λ_B, λ_G, λ_R, λ_{IR} represents the digital light signals within the visible/infrared spectrum and the wavelength arrows indicate their absorption depths during operation.

General purposes LEDs are used as light sources in two different ways: as digital signals (input channels) and as background lighting. The digital signals are impinged on the front side of the sensor. The background lighting is either at the back or at the front side. The intensity of the signal sources is very low compared to the background intensity. Different wavelength signal sources were used: violet (400nm), blue (470nm), green (524nm), red (626 nm), near-infrared (700 nm) and infrared (850 nm). For background lighting, 390 nm and 400 nm wavelengths were applied in a continuous and steady flux and different intensities. To change the background intensity different currents were used to drive the LED ($0<I_{LED}<30$ mA). When the background side changes the digital signals are sampled. The sensor is electrically biased at -8V.

BACKGROUND CONTROLLED LIGHT FILTERING EFFECTS

Spectral photocurrent measurements were accomplished with a monochromator in 10 nm steps from 400 nm to 800nm wavelengths. Spectral photocurrent results are presented in Figure 2 under front and back irradiation.

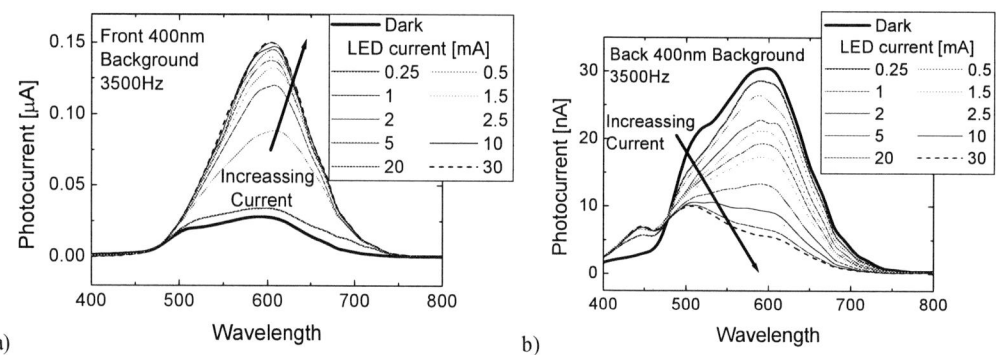

Figure 2 Spectral photocurrent with: a) front and b) back 400 nm background lighting.

The experimental results of Figure 2a show the photocurrent increases in the 470 nm-750 nm bandwidth. There is a significant increase just by the presence of the violet light; the high increase from no LED current to 0.5mA is outstanding when compared to the increase from 0.5 mA to 30 mA. In Figure 2b the thick black curve is the same from the previous figure and represents the dark level ($I_{LED}=0$). With the increase of the LED current the photocurrent in the 470 nm-800 nm bandwidth gradually decreases and there is an almost steady increase of the photocurrent in the 400 nm-470 nm bandwidth. The photocurrent gain is defined as the ratio between the photocurrent output and the value of the photocurrent when there is no background lighting. This gain is shown in Figure 3.

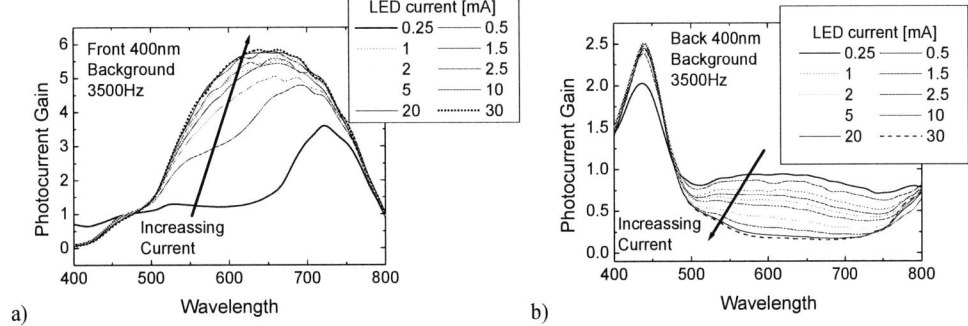

Figure 3 Photocurrent gain when background light is at the a) front and b) back side of the device.

Results show that the spectral gain shown in Figure 3a, reduces the short wavelengths (<470nm) and increases the higher wavelengths when the front side is illuminated. The opposite happens when the background lighting is at the back side (Figure 3b). Here the short wavelengths gain increase while the long ones decrease. This behavior can be used to build selective filters, where the gain of the short and long pass wavelengths is controlled by optical bias at either back or front sides. From the comparison of both a notch filter around 500 nm is also perceived.

VISIBLE/ INFRARED TUNING

Several monochromatic pulsed lights separately (850 nm, 697 nm, 626 nm, 524 nm, 470 nm, 400nm; input channels) or combined (MUX signal) illuminated the device at 12000 bps. Steady state 390 nm bias at different intensities due to different LED input currents ($0 < I_{LED} < 30$ mA) were superimposed separately from both sides and the photocurrent was measured.

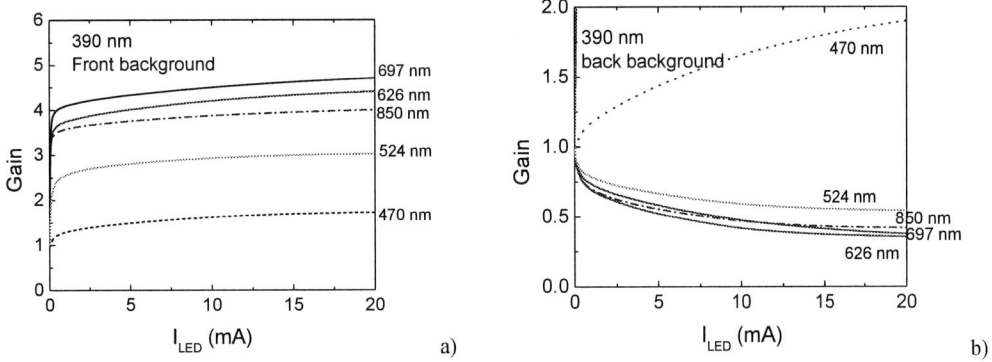

Figure 4 a) Front and b) back optical gain at λ=390 nm irradiation and different input wavelengths.

For each individual channel the photocurrent was normalized to its value without irradiation (dark) and the photocurrent gain determined. Figure 4 displays the different gain as a function of the drive currents of the lighting LED under front and back irradiation. Results show that the gain depends mainly on the channel wavelength and slightly on the lighting intensity [4]. Even across narrow bandwidths, the photocurrent gains are quite different. This nonlinearity allows the identification of the different input channels in the visible/infrared ranges. To exemplify, in Figure 5a, the gain of the 850 nm input channel, under front irradiation and different intensities, is displayed. In Figure 5b the MUX signals due to the combination of the 850 nm, 697 nm, 626 nm and 524 nm input channels are presented. At the top the signals used to drive the input channels are shown to guide the eyes into the *on/off* states.

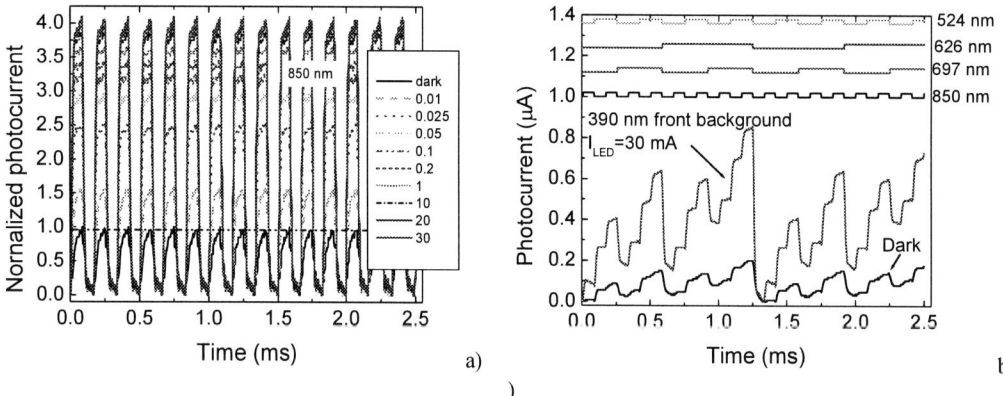

Figure 5   a) Optical gain at 390 nm front irradiation and different intensities. b) Combined polychromatic signal with and without 390 nm front irradiation and different bit sequences.

Results confirm that, even under transient conditions, the input channels present different gains, depending on their wavelengths [5]. This nonlinearity, due to capacitive effect at the internal interface [4, 5] allows channel recognition and so, a multiplexed information, due to several visible/infrared signals encoded in the same optical transmission path, can be decoded through a simple algorithm [5] that takes into account the channel gains and the illumination side (see Figure 4). Under front irradiation the sensor sensitivity to long wavelengths is enhanced and presents different sensitivities in a narrow red/infrared bandwidth when compared without background. The combination of the four channels under irradiation points out to the presence of all the possible sixteen (2^4) *on/off* states, clearly observed in Figure 5b. Here, each level is ordered by the correspondent gains in a 4 bit binary code [X_{697}, X_{626}, X_{850}, X_{524}] with X=1 if the channel is *on* and X=0 if it is *off*.The functional principle is based on the adjustable penetration depths of the photons into the front and back diodes (Figure 1) which is linked to their absorption coefficient in the intrinsic front and back collection areas. Front irradiation is strongly absorbed at the beginning of the front diode and, due to the self-bias effect, increases the electric field at the back diode where the red/infrared incoming photons are absorbed accordingly their penetration depths, and so to their wavelengths, resulting in an increased collection. Under back irradiation the electric field decreases mainly at the i-n back interface quenching the red/infrared signals and enhancing the blue /violet ones. Depending on the illumination side and intensity, the device sensitivity is tailored and shifted toward the long or short wavelength ranges leading to linearly profiled collection areas. So, by switching between front and back irradiation the photonic function is modified from a long- to a short-pass wavelength filter allowing, alternately selecting the red/infrared or the blue/violet channels, and making the bridge between the visible and the infrared regions. The green region is recognized through the use of both filters.

OPTOELECTRONIC MODEL

Based on the experimental results and device configuration an optoelectronic model, made out of a short- and a long-pass filter was developed [5] and upgraded to include several input channels. In Figure 6a the *ac* equivalent circuit and the block diagram of the optoelectronic state model are displayed. In Figure 6b the linearized state equations are shown. The use of amplifying elements (Q_1, Q_2), with resistors (R_1, R_2) and capacitors (C_1, C_2) in their feedback loops, synthesize the desired filter characteristics. $\alpha_{IR,R,G,B,V}$ are the observed experimental gain in the analyzed spectral regions.

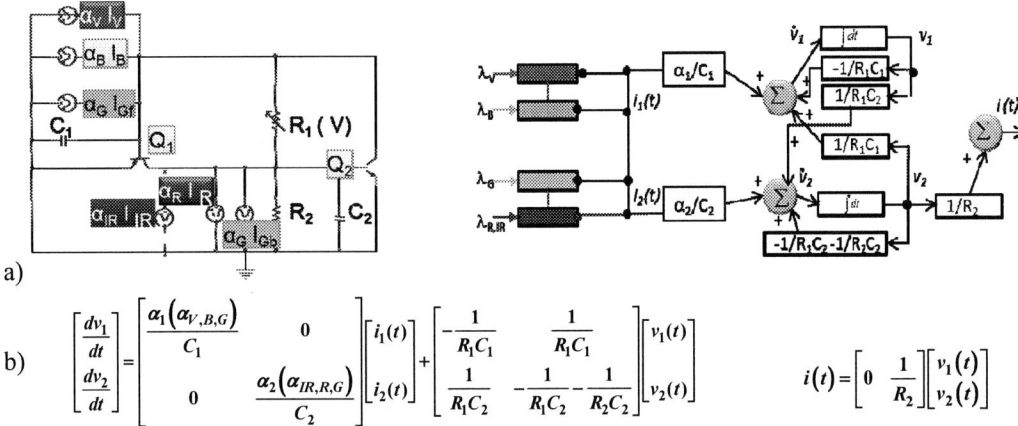

Figure 6 a) ac equivalent circuit and block diagram. b) Linearized state equations.

The input signals, $\lambda_{IR,R,G,B,V}$ model the input channels and $i(t)$ the output signal. The amplifying elements, α_1 and α_2 are linear combinations of the optical gains of each impinging channel, respectively into the front and back phototransistors and provide gain ($\alpha > 1$) if needed or attenuate ($\alpha < 1$) unwanted wavelengths. The control matrix takes into account the enhancement or quenching of the channels (Figure 4) due to the steady state irradiation. Under front irradiation: $\alpha_2 \gg \alpha_1$ and under back irradiation $\alpha_1 \gg \alpha_2$. This affects the reverse photo capacitances, ($\alpha_{1,2}/C_{1,2}$) that determine the influence of the system input on the state change (control matrix). A graphics user interface (GUI) computer program was designed and programmed within the MATLAB® programming language, to ease the task of numerical simulation. This interface allows selecting model parameters, along with the plotting of the bit signals, simulated and experimental photocurrent results. This is an upgraded version of the linearized differential equation model [5], using as solver one of two alternative algorithms available in MATLAB®: a one-step solver based on an explicit Runge-Kutta 4, 5 formula and an implementation of the trapezoidal rule using a "free" interpolant. To simulate the input channels we have use the individual magnitude of each input channel without background lighting, its bit sequence and the corresponding gain at the simulated background intensity (see Figures 4 and 5). Figure 7 present results of a numerical simulation with and without front λ=390 nm irradiation,

75

and different bit sequences (on top of the figures). A good fitting between experimental and simulated results was achieved.

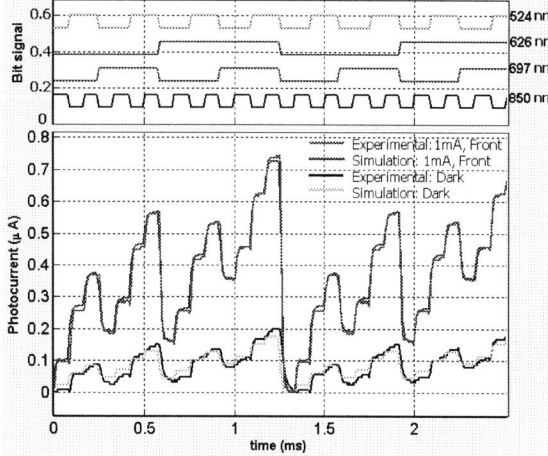

Figure 7 Simulation example of the MUX signal with and without front λ=390 nm irradiation.

The plots show the ability of the presented model to simulate the sensitivity behavior of the proposed system in the visible/infrared spectral ranges. A similar numerical simulation methodology was applied to the individual involved wavelengths, namely 524 nm, 626 nm, 697 nm and 850 nm, (not shown) with equally supportive results. The opto-electrical model with light biasing control has proven to be an adequate tool to design SiC multilayer filters. Furthermore, this model allows for extracting theoretical parameters (internal resistors and capacitors) by fitting the model to the measured data.

CONCLUSIONS

An increased sensitivity in a SiC pi'npin device in the VIS-NIR range under UV light was experimentally and theoretically demonstrated. Results show that under front 390 nm irradiation the sensor sensitivity was enhanced in the red/infrared ranges leading to linearly profiled collection areas that allow the incoming wavelength recognition. An optoelecronic model was presented to explain the observed data and to allow decoding a multiplexed information in the visible/infrared range.

ACKNOWLEDGEMENTS

This work was supported by FCT (CTS multi annual funding) through the PIDDAC Program funds and PTDC/EEA-ELC/111854/2009 and PTDC/EEA-ELC/120539/2010.

REFERENCES

[1]   S. Schmid, G. Corbellini, S. Mangold, and T. Gross. An LED-to-LED Visible Light Communication system with software-based synchronization. In Optical Wireless Communication. Globecom Workshops (GC Wkshps), 2012 IEEE, pp. 1264-1268 (2012).

[2]   G. Corbellini, S. Schmid, S. Mangold, T. R. Gross, and A. Mkrtchyan, "LED-to LED Visible Light Communication for Mobile Applications,"in Demo at ACM SIGGRAPH MOBILE 2012, 2012.

[3]   S. Mangold, "Toys communicating with LEDs: Enabling Toy Cars Interaction," in Demo at Consumer Communications and Networking Conference, CCNC, IEEE, 2012.

[4]   M. Vieira, P. Louro, M. Fernandes, M. A. Vieira, A. Fantoni and J. Costa "Advances in Photodiodes", InTech, Chap.19, pp:403-425 (2011).

[5] M. A. Vieira, M. Vieira, J. Costa, P. Louro, M. Fernandes, A. Fantoni "Double pin Photodiodes with two Optical Gate Connections for Light Triggering: A capacitive two-phototransistor model" Sensors & Transducers Journal, 9, Special Issue, 2011, pp.96-120.

Mater. Res. Soc. Symp. Proc. Vol. 1666 © 2014 Materials Research Society
DOI: 10.1557/opl.2014.720

Optical Characterization of Si Nanowires:
Dependence with Substrate Orientation and Light Polarization

Juan A. Badán[1], Ricardo E. Marotti[1], Enrique A. Dalchiele[1], Daniel Ariosa[1], Francisco Martín[2], Dietmar Leinen[2], José R. Ramos-Barrado[2].

[1]Instituto de Física & CINQUIFIMA, Facultad de Ingeniería, Universidad de la República, Julio H. Reissig 565, CC 30, CP 11000, Montevideo, Uruguay.

[2]Lab. de Materiales y Superficies (Unidad Asociada al CSIC), Dptos. de Física Aplicada & Ingeniería Química, Universidad de Málaga, Campus de Teatinos s/n, E29071 Málaga, Spain.

ABSTRACT

Optical properties of Si nanowire arrays (SiNWs) prepared on p-doped Si(111) and Si(100) substrates are studied. The SiNWs were synthesized by self-assembly electroless metal deposition nanoelectrochemistry in an ionic silver HF solution through selective etching. Total reflectance (R_t) and total diffuse reflectance (R_{dt}) of SiNWs change drastically in comparison to polished Si. To understand these changes diffuse reflectance (R_d) with polarized incident light was studied. For samples prepared on Si(111), the wavelength integrated R_d (wIRd) shows maxima at certain angle of incidence θ and it does not depend on light polarization. Moreover, R_{dt} of SiNWs prepared on Si(111) can be modeled as an ensemble of diffuse reflectors. For samples prepared on Si(100) wIRd increases with θ, being greater when the light electric field is parallel to the plane of incidence. Also, R_d spectra show structures due to interference effects. For these reasons SiNWs prepared on Si(100) can be considered as a thin film whose refractive index depends on light polarization.

INTRODUCTION

Silicon nanowire arrays (SiNWs) have important antireflective properties [1]. One reason is their gradual variation of effective refractive index from air to substrate [2,3]. Another one is light trapping by multiple scattering events [2,4], which increases the probability of light absorption [4]. For these reasons their main application is for light enhanced absorption in Si photovoltaic solar cells [5,6]. Another advantage for this application is the possibility of decoupling light absorption and charge carrier collection into orthogonal directions. This condition is satisfied with a junction in the radial direction, and the nanowire (NW) axis parallel to the incident light direction [7,8], e. g. normal incidence of light when nanowires (NWs) are perpendicular to the substrate plane. In this way, the solar cell requirements on the minority carrier diffusion length of the absorber material are reduced significantly in comparison to planar geometry [7-10].

Among the many methods developed for the growth of SiNWs, self-assembly electroless metal deposition (SAEMD) nanoelectrochemistry in an ionic silver HF solution through selective etching is a simple way to prepare SiNWs arrays [11,12]. This method has several advantages with respect to other SiNWs preparation techniques. First, since the as prepared SiNWs are an integral part of the Si wafer substrate, they provide a direct 1D and uninterrupted pathway for charge transport to the substrate. In addition the electrical properties of SiNWs are directly inherited from the bulk Si substrate, thus there is no need of further doping for conductivity control. In this item an important fact is that orientation of NWs depends on orientation of bulk crystalline Si (c-Si) substrate [13]. Finally, due to the surface roughness of the SiNWs, they are almost non reflective [14, 15]. In this paper this very small optical reflectance and its relation with c-Si orientation is studied for SiNWs grown from both (111) and (100) Si wafers. Moreover, the influence of light polarization with substrate orientation is also investigated.

EXPERIMENTAL

Single-side polished Si(111) wafer chips (WCh), 500 μm thick, p-doped (50 - 100 Ωcm) and both-side polished Si(100) WCh, 280 μm thick, p-doped (1.0 - 5.0 Ωcm), with areas of ~ 1 cm², were first washed in boiling acetone for 10 min and subsequently in isopropanol at room temperature (RT), with sonication for 5 min. They were oxidized in $H_2O_2/HCl/H_2O$ (2:1:8) at 353 K for 15 min to remove any trace of heavy metals and organic species, rinsed copiously with de-ionized water, etched with 10 % aqueous (aq) HF for 10 min, rinsed with water again, dried under a stream of N_2 and immediately used as a substrate in the SiNWs growth process. The SiNWs have been synthesized by SAEMD nanoelectrochemistry on the Si WCh in an ionic silver HF solution (sol) through selective etching [14,17]. The cleaned Si WCh were immersed into an aq $HF/AgNO_3$ sol for 15 - 30 min at RT (the concentration of HF and $AgNO_3$ here are chosen to be 5.0 M and 0.04 M, respectively). The length of SiNWs could be effectively controlled through tuning the treatment time; in the present case one hour treatment has been applied. After the treatment, the as-synthesized samples were rinsed copiously in deionized water and dried at RT. Then, the SiNWs were dipped in 30 wt. % HNO_3 aq sol for 60 s and repeated for several times to remove all residual Ag from the SiNWs surfaces. After the HNO_3 bath no Ag peaks appeared in the Energy Dispersive X-ray Spectroscopy (results not shown). Scanning Electron Microscopy (SEM) pictures were obtained with a JEOL JSM-5410 apparatus.

Total reflectance R_t (diffuse + specular) and total diffuse reflectance R_{dt} were measured by an integrating sphere Ocean Optics (OO) ISP-REF and an OO S2000 spectrometer. A lambertian surface of spectralon OO WS-1 SL White Reflectance Standard was used as reference. Because specular reflectance of SiNWs is very low, diffuse reflectance R_d was measured varying the angle of incidence θ. The experimental setup is shown in figure 1. Light from an OO HL2000 halogen lamp was collimated by an OO 74Vis Collimating Lens.

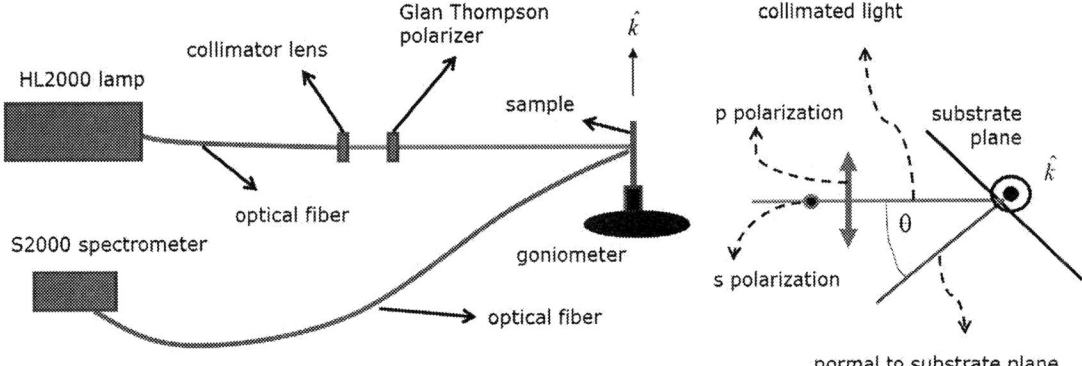

Figure 1. Experimental setup for R_d measurement with polarized light. Schematic to the right describes the *p* and *s* light polarizations and angle of incidence θ.

R_d was detected by an optic fiber with 100 µm diameter coupled to the OO S2000 (acting as its entrance slit). The other end of the fiber was arranged close to the sample, mounted on a goniometer which allows varying the angle of incidence θ, in a quasi-Littrow configuration. The axis of rotation and the plane of the substrate are parallel to vertical direction $\hat{k}$ (see figure 1), which is perpendicular to the plane of incidence. A Glan Thompson polarizer was arranged after the collimating lens. The two distinct polarizations (see schematics to the right of figure 1) were *p* (light electric field parallel to the plane of incidence) and *s* (perpendicular to the plane of incidence).

RESULTS AND DISCUSSION

Figure 2 shows images of SiNWs prepared on Si(111) and Si(100) c-Si substrates, to be called Si(111)NWs and Si(100)NWs, respectively. The Si(111)NWs have a brown-yellow appearance to the naked eye, with a milky shimmering like brightness when looked sidewise. Meanwhile the Si(100)NWs are dull black (see figure 2a). The SEM images (figures 2b and 2c) show that the NWs have large aspect ratio (150 nm to 200 nm diameters, with 5 µm to 7 µm length). In both cases there is agglomeration of NWs at their tops ends [3,6]. But while the Si(111)NWs have a disordered slanting morphology (figure 2b), the Si(100)NWs have an ordered one, with NWs normal to the substrate (figure 2c). This agrees with the fact that in both cases the NWs grow along <100> directions [13].

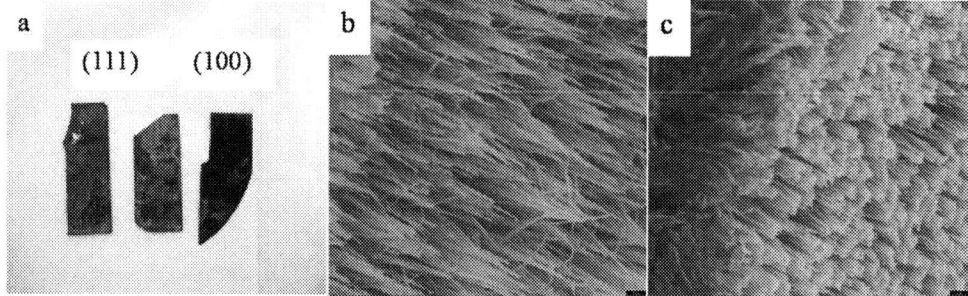

Figure 2. a) Image of Si(111)NWs (left and center) and Si(100)NWs (right). SEM image of b) Si(111)NWs and c) Si(100)NWs. Black bar in b) and c) low right angle is 2 µm.

Figure 3a shows R_t and R_{dt} of SiNWs and c-Si. R_t of polished c-Si decreases with wavelength and shows a peak (~ 380 nm) which corresponds to $\Gamma_{25'}$-Γ_{15} direct transition [18]. Meanwhile, R_{dt} c-Si shows a step (between 700 nm – 800 nm) which corresponds to $\Gamma_{25'}$-L_{1C} indirect transition [19]. Spectra shapes change drastically for SiNWs. R_t of SiNWs samples are, in both cases, smaller than R_{dt} of polished c-Si, and the peak corresponding to $\Gamma_{25'}$-Γ_{15} disappears. Besides, the step is less abrupt for Si(100)NWs and disappears for Si(111)NWs. For Si(111)NWs the smooth monotonic increase of R_{dt} with wavelength gives them their brown-yellow appearance. Although both R_{dt} and R_t have similar spectral shape, the first one is smaller than the second one because the specular reflectance is still important in this case. However, for Si(100)NWs both R_{dt} and R_t are almost equal (vanishing specular reflectance). Moreover, they are much smaller than for Si(111)NWs, in agreement with their dull black appearance.

The changes in Si(111)NWs R_{dt} can be modeled as an ensemble of diffuse reflectors [4]. According to this model, the R_{dt} of a mat (R_{dtMat}) of diffusing NWs, neglecting transmittance, is:

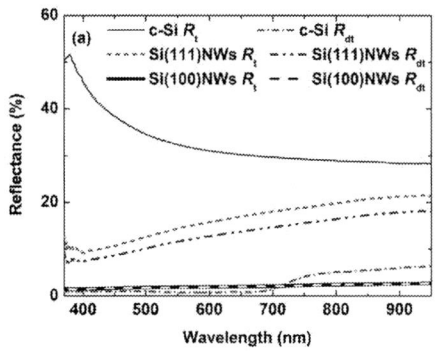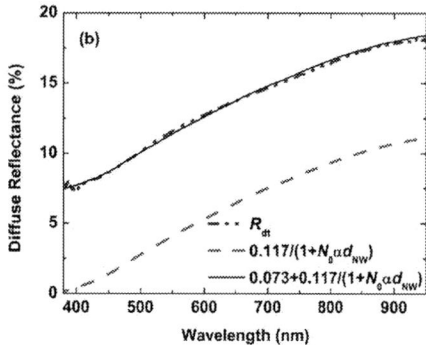

Figure 3. a) R_t and R_{dt} for polished c-Si, and SiNWs. For Si(100)NWs both curves are indistinguishable. b) Measured R_{dt} for Si(111)NWs and model.

$$R_{dtMat} = \left[1 + N_0 \alpha d_{NW}\right]^{-1} \tag{1}$$

where α is the absorption coefficient of Si (obtained from tabulated data [20]), d_{NW} is an absorption length, expected to be roughly the NW diameter (assuming all NWs are lying on their side) [4], N_0 is a parameter of $P_N = N_0^{-1} \exp(-N/N_0)$, an exponential distribution assumed for the probability that a photon is reflected out of the mat after N scattering events [4]. $N_0 d_{NW}$ depends on NW diameter and length [2]. Using $N_0 d_{NW} = 3 \times 10^{-4}$ cm corresponding to 4 – 7 μm length and 130 nm – 150 nm diameter NW [2] R_{dtMat} almost vanishes into the UV, and is 0.97 at 950 nm (where $\alpha N_0 d_{NW} = 0.03$). So the following corrections are needed to adjust the measured R_{dt}:

$$R_{dt} \sim R_{NS} + (R_{dt@950nm} - R_{NS})R_{dtMat}/R_{dtMat@950nm} \tag{2}$$

where $R_{NS} \sim R_{dt@UV} = 0.073$ is an additive correction [4] and $(R_{dt@950nm} - R_{NS})/R_{dtMat@950nm}$ is a factor in R_{dtMat} which allows adjusting the IR value ($R_{dt@950nm} = 0.18$). Equation (2) fits very well the R_{dt} of Si(111)NWs (see figure 3b).

Figure 4 shows R_d of SiNWs, for s-polarized light and for different θ values. For Si(100)NWs the reflectance is much lower than for Si(111)NWs. To study the θ dependence, the wavelength integrated diffuse reflectance (wIRd) between 380 nm and 900 nm were calculated. The results are shown in figure 5. Although the wIRd increases monotonically with θ for Si(100)NWs, there is a relative maximum between 40° and 50° for Si(111)NWs. Both results can be interpreted considering that NWs grow along <100> directions. Indeed, for Si(111)NWs the sample was oriented such that the substrate [001] direction lays in the plane defined by the $\hat{k}$ vector of figure 1 and the normal to substrate surface (orientation obtained from X-Ray Diffraction pole figure, results not shown). For $\theta \sim \pm 39.2°$, light approaches normally to [100] or [010] directions, corresponding to the maximal NWs geometrical cross section, i. e. maximum backscatter reflectance. This θ is close to the maxima in figure 5a within experimental θ step, and is responsible of the shimmering like brightness of these samples. The different maximum intensities (for $\theta > 0$ or < 0) is due to the random grow of NWs in any of <100> directions [13]. There is not a clear difference between wIRd for polarization p and wIRd for polarization s with respect to θ.

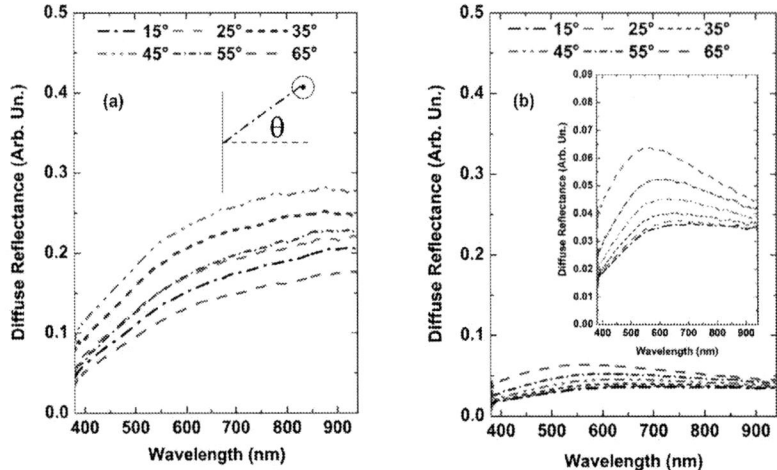

Figure 4. R_d for *s* polarization of a) Si(111)NWs (schematics shows polarization) and b) Si(100)NWs varying θ (inset is zoomed vertical axis).

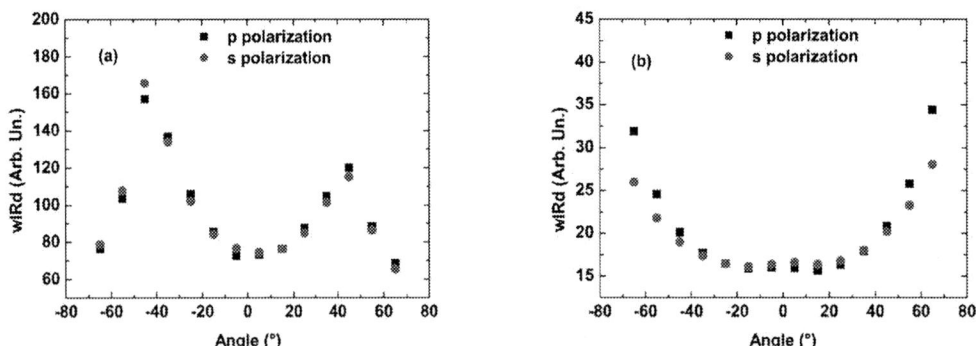

Figure 5. wIRd, area below R_d of figure 4: a) Si(111)NWs and b) Si(100)NWs against θ.

For Si(100)NWs (figure 5b) wIRd increases monotonically with θ, increasing angle between NW axis and light direction. Also, wIRd increase for *p*-polarization is greater than for *s*-polarization. This is coherent with a polarization dependent effective refractive index *n* [21]:

$$n_p(\theta) = \left(n_s^2 + \left(n_p(90^0)^2 - n_s^2 \right) \sin^2 \theta \right)^{1/2} \tag{3}$$

where n_p (n_s) is the effective refractive index for *p* (*s*) polarization. Reflected light increases for larger change of *n* among air and that of the SiNWs layer. For *s* polarization the electric field is always perpendicular to NW axis (figure 1), if they are grown normal to substrate (figure 2c) ([100] direction). While for *p* polarization, the electric field has a parallel component along these axes, which increases with θ. As the SiNWs effective *n* is higher along this direction, the wIRd at large θ must be higher for *p* than for *s* polarization: $n_p(0^\circ) = n_s$, $n_p(90^\circ) > n_s$. Note that the

82

difference in wIRd between both polarizations in figure 5b (at least for larger θ) is clearly bigger than their difference in the whole span of figure 5a. This can be understood because of the more disordered nature of Si(111)NWs (see figure 2). This disorder is responsible of the strong light scattering in the NWs that allows writing equation (2) that governs the spectral shape of Si(111)NWs R_{dt}.

Finally, figure 4b inset shows that (especially for higher θ), the R_d for Si(100)NWs has a structure that does not appear so clear in Si(111)NWs (figure 4a). This structure might be due to interference effects [2,3]. It reflects the ordered nature of these SiNWs array (figure 2b). This behavior and the dependence with light polarization allow considering Si(100)NWs as a layer with an effective refractive index.

CONCLUSIONS

The optical properties of Si nanowire arrays (SiNWs) prepared into Si(111) and Si(100) crystalline substrate (c-Si) by Electroless Metal Deposition, were studied. Those prepared onto Si(111) (Si(111)NWs) have brown-yellow aspect while those prepared onto Si(100) (Si(100)NWs) have a dull black appearance to the naked eye. The total reflectance (R_t) and total diffuse reflectance (R_{dt}) of SiNWs spectral shapes change drastically in relation to c-Si. SiNWs has smaller R_t than polished c-Si. R_{dt} for Si(111)NWs can be modeled as an ensemble of diffuse reflectors. Besides, diffuse reflectance R_d for Si(111)NWs does not depend strongly on light polarization, but the wavelength integrated R_d (wIRd) shows maxima at certain angles of incidence θ. They can be explained if nanowires grow along <100> directions. For Si(100)NWs wIRd increases monotonically with θ, being wIRd for p-polarized incident light greater than for s-polarized light. Also, R_d spectra show structures which correspond to interference effects. Thus Si(100)NWs can be considered as a thin film whose refractive index depends on the light polarization.

ACKNOWLEDGMENTS

CSIC (Comisión Sectorial de Investigación Científica) of the Universidad de la República; ANII (Agencia Nacional de Investigación e Innovación); PEDECIBA – Física, in Uruguay. Junta de Andalucía, Project P07-FQM-02573 in Spain.

REFERENCES

1. K. Peng, X. Wang, and S.-T. Lee, Appl. Phys. Lett. **92**, 163103 (2008).
2. A. Convertino, M. Cuscuna, and F. Martelli, Nanotechnology **21**, 355701 (2010).
3. W. Q. Xie, J. I. Oh, and W. Z. Shen, Nanotechnology **22**, 065704 (2011).
4. R. A. Street, W. S. Wong, and C. Paulson, Nano Letters **9**, 3494 (2009).
5. X. X. Lin, X. Hua, Z.G. Huang, and W. Z. Shen, Nanotechnology **24**, 235402 (2013).
6. Y. Jiang, R. Qin, M. Li, G. Wang, H. Ma, and F. Chang, Mater. Sci. Semicon. Proc. **17**, 81 (2014).
7. J. M. Spurgeon, H.A. Atwater, and N.S. Lewis, J. Phys. Chem. C **112**, 6186 (2008).
8. B. M. Kayes, H. A. Atwater, N. S. Lewis, J. Appl. Phys. **97** 114302 (2005).
9. R. Tena-Zaera, M.A. Ryan, A. Katty, G. Hodes, S. Bastide, and C. Levy-Clemént, C. R. Chimie **9**, 717 (2006).

10. J. B. Baxter and E. S. Aydil, Appl. Phys. Lett. **86**, 053114 (2005).
11. K.-Q. Peng, Y.-J. Yan, S.-P. Gao, and J. Zhu, Adv. Mater. **14**, 1164 (2002).
12. T. Qiu, X. L. Wu, Y. F. Mei, G. J. Wan, P.K. Chu, and G. G. Siu, J. Cryst Growth **227**, 143 (2005).
13. S.-L. Wu, T. Zhang, R.-T Zheng, and G.-A. Cheng, Appl. Surf. Sci. **258**, 9792 (2012).
14. H. M. Branz, V. E. Yost, S. Ward, K. M. Jones, B. To, and P. Stradins, Appl. Phys. Lett. 94, 231121 (2009).
15. J. Oh, H.-C. Yuan, and H. M. Branz, Nature Nanotechnology, **7**, 743 (2012).
16. E. A. Dalchiele, F. Martín, D. Leinen, R. E. Marotti, and J. R. Ramos-Barrado, Thin Solid Films **518**, 1804 (2010).
17. E. A. Dalchiele, F. Martín, D. Leinen, R. E. Marotti, and J. R. Ramos-Barrado, J. Electrochem. Soc. **156**, K77 (2009).
18. M. Nelkomsky and R. Braunstein, Phys. Rev. B **5**, 497 (1972).
19. R. A. Forman, W. R. Thurber, D. E. Aspnes, Solid State Commun. **14**, 1007 (1974).
20. M. A: Green and M. Keevers, Prog. Photovolt: Res. Appl. **3**, 189 (2007).
21. C. J. Oton, Z. Gaburro, M. Ghulinyan, L. Pancheri, P. Bettotti, L. Dal Negro, and L. Pavesi, Appl. Phy. Lett. **81**, 4920 (2002).

Femtosecond laser materials processing of a-Si:H below the ablation threshold

B. Soleymanzadeh[1], W. Beyer[2,3], F. Luekermann[1], P. Prunici[4], W. Pfeiffer[1], and H. Stiebig[1,5]

1. Molecular and Surface Physics, Bielefeld University, D-33615 Bielefeld, Germany
2. Institut für Silizium-Photovoltaik, HZB, Kekuléstrasse 5, D-12489 Berlin, Germany
3. IEK5-Photovoltaik, Forschungszentrum Jülich GmbH, D-52425 Jülich, Germany
4. Solayer GmbH, Sachsenallee 28, D-01723 Kesselsdorf, Germany
5. Institut für Innovationstransfer an der Universität Bielefeld, Universitätsstr. 25, D-33615 Bielefeld, Germany

ABSTRACT

Laser processing of thin-film silicon is a promising approach for the realization of polycrystalline silicon for large area electronics and solar cell applications. In this study we investigate the material modification of amorphous hydrogenated silicon (a-Si:H) with different hydrogen content (30%, 13% and <1%) by means of femtosecond (fs) laser pulses. Depending on the peak fluence applied, hydrogen diffusion/effusion, layer crystallization or material ablation can be achieved. Despite the low absorption coefficient of a-Si:H at the center wavelength of an amplified Titanium Sapphire laser at 790 nm a high local energy deposition close to the surface of the a-Si:H layer is observed, which can be attributed to a nonlinear absorption process.

INTRODUCTION

Material modification of amorphous silicon (a-Si:H) by solid phase crystallization [1,2], electron-beam crystallization [3] or laser crystallization [4] for solar cell application has been successfully demonstrated over recent years. In the field of laser materials processing the applied laser pulse duration ranges from femtoseconds up to continuous wave (CW) with different wavelengths. Since the thermally introduced effects decrease with shorter pulse duration, femto second (fs) laser material processing enables new applications in the field of a-Si:H [5-7]. In this study we investigate the material modification of a-Si:H deposited by plasma enhanced chemical vapor deposition (PECVD) by means of femtosecond laser pulses. Since an amplified Titanium Sapphire laser with a center wavelength of 790 nm is used, a two photon absorption process takes place. a-Si:H films with different hydrogen content (30%, 13% and <1%) were studied to investigate the influence of the hydrogen content of the films on the ablation threshold.

EXPERIMENT

The different hydrogen content of the intrinsic a-Si:H layers was achieved by varying the deposition temperature of the PECVD process (25°C, 200°C and 520°C). In particular a-Si:H layers with a thickness of 50 nm and >290 nm were used for laser modification experiments. Single 30 fs long laser pulses of an amplified Titanium Sapphire laser (1 kHz repetition rate, 790 nm center wavelength) were attenuated and focused to an e^{-2}-spot width of 420 µm. The peak fluence range used for the irradiation experiments was 20 mJcm^{-2} to 120 mJcm^{-2}. The single shot modified areas were characterized by optical microscopy, imaging ellipsometry at 658 nm, scanning electron microscopy (SEM), and micro-Raman spectroscopy. The spot diameter of the Raman laser is in the range of 1 to 2 μm. This allows for recording of spatial resolved Raman signals from single laser pulse modified regions by scanning the focus across the region of interest. The measured intensity profile of the focus in combination with microscopy allows analyzing the fluence dependency of the material modification by recording spatially resolved signals across individual spots. Qualitative depth information of the material modification is obtained from micro-Raman spectroscopy using different excitation wavelengths (473 nm and 633 nm) with different penetration depth profiles. Beside the evaluated crystalline fraction from spectroscopic measurements the Raman signals associated with Si-H vibrational modes (2000-2100 cm^{-1} Stokes shift) were used for the characterization of the hydrogen content.

RESULTS AND DISCUSSION

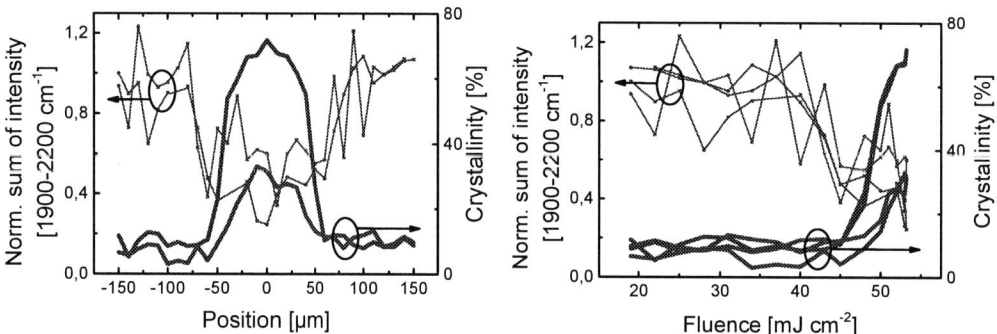

Figure 1. Crystalline fraction and hydrogen content as a function of the position (left) and as a function of the local incident fluence (right) of an a-Si:H film (hydrogen content = 13%, film thickness = 305 nm) locally modified by a single fs laser pulse with a peak fluence of 53 mJcm^{-2}. Since the line-scan is performed through the modified spot area, the graphs showing the data (e.g. hydrogen content, crystallization) as a function of the fluence contain always two data sets.

Fig. 1 shows the Raman crystallinity (bold solid line) and the normalized hydrogen content (dotted line) as a function of the local position (left) and the fluence (right) of an a-Si:H film (hydrogen content = 13%) modified by a single fs laser pulse of 53 mJcm^{-2}. Due to the Gaussian intensity distribution of the laser spot, a maximum of the crystalline volume fraction and a minimum of the hydrogen content is observed at the center of the focus (fig. 1, left), which correlates with the position x = 0. Whilst the crystallinity retrieved from micro-Raman spectroscopy performed at 473 nm und 633 nm excitation wavelengths agrees well for distances of more than 50 μm from the laser spot center, clear differences between both signals are obvious in the spot center. Depending on the excitation wavelength used, a Raman crystallinity of 70% and 30% was determined. Due to the different penetration depth Raman signals for excitation light at 472 nm predominantly probe the surface region, whereas for excitation light at 633 nm the signal originates from the whole layer. From the different Raman signals we conclude that fs-laser material processing at fluences of above 50 mJcm^{-2} induces the crystallization of a relatively thin surface layer. This indicates a strong non-linear absorption of the fs laser pulse, since a-Si:H has a low absorption coefficient at the central fs laser wavelength at 790 nm and a linear absorption process would thus excite the whole layer.

The normalized amplitude of the measured Si-H stretching mode (ratio of the local amplitude divided by the amplitude measured at an untreated area) significantly decreases at the surface by around 75% and in the bulk of the layer by 50%. Taking into account a Gaussian intensity distribution of the laser beam, the hydrogen content and the Raman crystallinity can be determined as a function of the local incident laser fluence. Fig. 1 (right) reveals that the threshold fluence for the hydrogen effusion and crystallization is about 40 mJcm^{-2} and 50.0 mJcm^{-2}, respectively. Thus, before the crystallization process starts, the hydrogen concentration is already significantly reduced. At this point it has to be mentioned that the microstructure factor (ratio of the vibration mode at 2000 cm^{-1} divided by the sum of 2000 cm^{-1} and 2100 cm^{-1}) remains constant within the whole line scan. Similar threshold fluences are observed for material processing spots created with varying peak fluence, indicating that lateral heat diffusion does not affect the material processing for the present excitation conditions.

In order to assess the derived threshold value for hydrogen diffusion/effusion the deposited energy can be estimated by means of a non-linear absorption model and compared with data derived from thermal annealing experiments. Assuming a two photon absorption (TPA) coefficient of 1 nmW^{-1} [8] and a reflectivity of the a-Si:H layer of 63%, a deposited energy density of 2 kJcm^{-3} close to the surface region is derived. To benchmark this value thermal annealing experiments of co-deposited a-Si:H samples on c-Si substrates are conducted. After 5 min temperature treatment at 300°C a decrease of the hydrogen concentration is observed. Taking into account the layer thickness, specific heat capacitance, and the material density, the transferred heat energy is around 0.5 kJcm^{-3}. Considering the uncertainties of the non-linear absorption model and the fact that heat diffusion is completely neglected for the estimation of the local laser deposited energy density, both estimated energy densities agree well. This confirms that fs laser pulses deposit energy very efficiently in the material and that Si-H bond breaking occurs. Under steady state conditions the diffusion length of hydrogen in a-Si:H at 300 K is in the range of several nm and it is thus unlikely that hydrogen will leave the layer after fs-laser materials processing. Recombination of the released H to H_2 within small voids that might explain the reduced hydrogen concentration was ruled out by Raman spectroscopy at H_2 vibration at 3601 cm^{-1} and 4157 cm^{-1} [9]. No variation of the H_2 line is found in a line scan across any fs laser pulse treated area and, thus, an enhanced occupation of H_2 in small voids seems unlikely. Because of the non-linear absorption process a high local energy density is reached, resulting in local temperatures exceeding 1000 K for peak laser fluences higher than 40 mJcm^{-2}. Thus, an out-diffusion of H is more pronounced and can account for the reduced hydrogen related Raman signal.

Beside the Raman characterization also imaging ellipsometry at 658 nm was carried out. Fig. 2 exhibits the amplitude component Ψ and the phase difference Δ of the complex reflectance ratio as a function of the fluence measured by ellipsometry. Ψ and Δ show characteristic features at 40 mJcm^{-2} and 46 mJcm^{-2}, which correlate well with the thresholds for hydrogen diffusion/effusion and the crystallization (see fig. 1, right). Furthermore, the amplitude component Ψ slightly decreases from 27.8 to 27.3 between 30 mJcm^{-2} and 40 mJcm^{-2}. Surprisingly, this small variation in the amplitude component Ψ is accompanied by a color variation in the optical microscope image (not shown). Therefore, there exists an additional further characteristic threshold at about 30 mJcm^{-2}, indicating a change in the material properties, although microscopic properties probed by Raman spectroscopy remain unchanged. Further investigations are necessary to study the influence of this threshold on the structural or optoelectronic properties in more detail.

Finally the material ablation threshold of standard a-Si:H was determined. For a-Si:H with a hydrogen content of around 13%, a peak fluence exceeding 55 mJcm^{-2} is required for material ablation.

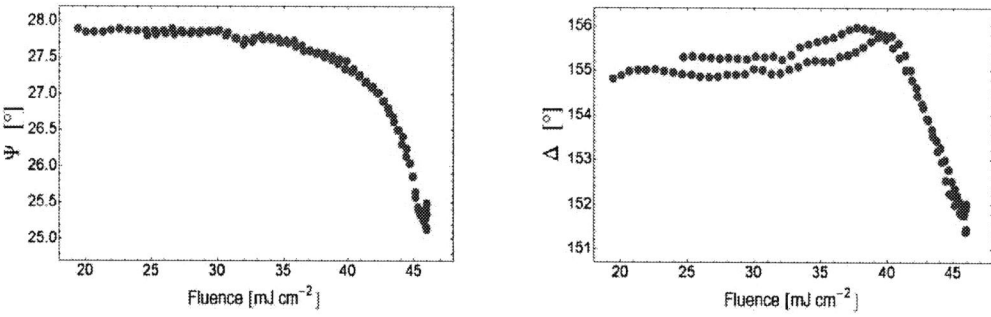

Figure 2. Ellipsometric parameters (Ψ and Δ) as a function of the fluence across a fs laser modified area of a standard a-Si:H layer. The applied peak laser fluence is 46 mJcm^{-2} as a function of the fluence across a fs laser modified area of a standard a-Si:H layer. The applied laser peak fluence was 46 mJcm^{-2}.

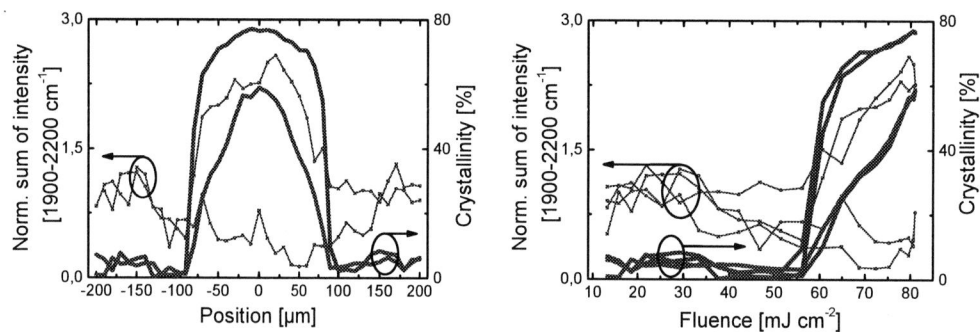

Figure 3. Cross section of the crystallinity and hydrogen content of a fs modified area (left) and crystallinity and normalized hydrogen content as a function of the fluence (right) of a-Si:H deposited at 520°C (hydrogen content < 1%, layer thickness = 490 nm) processed by a single shot fs laser pulse of 81 mJcm^{-2}.

Fig. 3 shows the crystallinity of a sample with a low hydrogen content (<1%) processed by a single shot fs laser pulse of 81 mJcm^{-2}. Caused by the higher peak fluence in comparison to the standard a-Si:H sample (fig. 1, left), the on-set of crystallization appears more separated from the focus center (fig. 3, left). In the focus center a crystallinity of around 75% and 60% is reached by performing Raman spectroscopy at 473 nm und 633 nm excitation wavelengths, respectively. In addition, the crystallization threshold is only slightly increased (fig. 3, right) in comparison to the standard material. Also for this sample, a higher Raman crystallinity is measured at the surface in comparison to the bulk, even for the highest local fluence. This indicates again that a high local energy deposition close to the surface of the a-Si layer is achieved in particular for fluences slightly above the crystallization threshold.

A strong energy deposition in a thin surface layer of a-Si:H by an ultra-short laser pulse at a center wavelength at 790 nm can only be achieved when non-linear absorption occurs. For non-linear absorption, the intensity within the a-Si:H layer can be described by the following differential equation (1) [8]:

$$\frac{dI}{dz} = \left(\frac{\alpha_0}{1 + \frac{I}{I_s}} + \beta I \right) I \qquad (1),$$

where I is the irradiance intensity, z denotes the penetration depth, α_0 is the linear absorption coefficient and I_s is the saturation intensity. β denotes the two photon absorption (TPA) coefficient and has a value of around $1 nm W^{-1}$ [8].

A further interesting feature shows the measured amplitude of the Si-H vibration mode ($2000-2100$ cm^{-1} Stokes shift) in particular at the surface. Even below the ablation threshold the surface Raman signal increases within the crystallized area by more than a factor of two in comparison to an untreated area. The higher amplitude of the Si-H vibration mode might be attributed to i) in-diffusion of H_2O, ii) chemical surface reactions during laser processing involving hydrogen containing compounds, iii) laser induced accumulation of hydrogen from the film at the surface or iv) variation of the Raman efficiency. To investigate this feature in more detail, we performed fs laser materials processing of undoped a-Si:H films deposited by e-beam evaporation. For the e-beam evaporated films no enhanced Si-H vibration mode at the surface was found. Therefore, the enhanced Si-H surface concentration of the fs laser modified PECVD sample is likely to originate from the bulk material itself. Considering the low bulk hydrogen concentration in the film (<1%), the continuous decrease of the bulk Si-H Raman signal (red dashed curve in Fig. 3) and the coincidence of the signal increase with the crystallization onset, we conclude that an enhanced Raman efficiency at 473 nm excitation is caused by modified surface properties.

Next, the ablation threshold of a-Si:H films with a low hydrogen concentration was studied. Laser pulses with peak fluences of about 90 $mJcm^{-2}$ are required for material removal. For peak fluences up to 115 $mJcm^{-2}$ holes with a depth of around 25 nm are created.

For a sample with an even larger hydrogen concentration ($T_{dep} = 25°C$, hydrogen concentration > 30%) material ablation was observed at a peak fluence $\cong 40$ $mJcm^{-2}$. Below this fluence neither a decrease of the hydrogen content nor a crystallization of the film surface is observed in the hydrogen rich film. In contrast, for samples with a low hydrogen concentration (a-Si:H prepared by high temperature PECVD growth or e-beam evaporation) material ablation requires a fluence of about twice the size i.e. 90 $mJcm^{-2}$. The crystallization and ablation thresholds for e-beam evaporated and high temperature deposited PECVD samples agree very well. All these observations indicate that the ablation threshold is mainly influenced by the hydrogen concentration. For peak fluences above the ablation threshold (50 $mJcm^{-2}$ - 75 $mJcm^{-2}$) the created holes have a depth of 50 nm only. A crystallization of a-Si:H at the bottom of the holes is detected for a peak fluence greater than 60 $mJcm^{-2}$. Considering that a certain material volume is removed by irradiation with a 60 $mJcm^{-2}$ laser pulse, the value for the on-set of crystallization in the created holes also has to be higher in comparison to the samples with a lower hydrogen concentration. Thus the thresholds for hydrogen release and crystallization do not significantly depend on the hydrogen concentration.

The complex behavior of fs laser material processed a-Si:H and the influence of the hydrogen content on the material modification can be qualitatively explained in the following model. For low hydrogen content, hydrogen is mobilized at 40 mJcm^{-2} incident fluence and recrystallization starts at 50 mJcm^{-2}. The Si-H bond dissociation has a strong influence on the ablation behavior. At medium and high hydrogen concentration the H release destabilizes the layer and ablation occurs at much lower fluences compared to films with a low hydrogen concentration. At this point is has to be mentioned that the ablation threshold also depends on the layer thickness. For thinner layers (thickness = 50 nm) the ablation threshold shifts to lower peak fluences and the whole layer is locally removed. This may well be attributed to the adhesion behavior of silicon at the glass substrate and accumulation of hydrogen at the silicon/glass interface and bubble induced material removal.

CONCLUSIONS

fs laser processing of a-Si:H has shown that the ablation threshold of a-Si:H strongly depends on the hydrogen content while the threshold for hydrogen diffusion/effusion and crystallization is in the range of 40 mJcm^{-2} and 50 mJcm^{-2}, respectively. Despite the low absorption coefficient of a-Si:H at 790 nm a high local energy deposition close to the surface of the a-Si layer is achieved via nonlinear absorption. Consequently, a distinct material modification (hydrogen diffusion/effusion, crystallization, ...) in a thin layer near the surface can be achieved making it possible to produce electronic circuits for large area electronics and to modify contact and intermediate layers for solar cell application.

ACKNOWLEDGEMENTS

Support by BMU (Globe-Si project, No. 0325446) is gratefully acknowledged.

REFERENCES

1. T. Matsuyama, N. Terada, T. Baba, T. Sawada, S. Tsuge, K. Wakisaka, S. Tsuda, Journal of Non-Crystalline Solids, 198–200 (1996) 940.
2. M. J. Keevers, T. L. Young, U. Schubert, M. A. Green, Proc. 22nd European Photovoltaic Solar Energy Conference, Milan, Italy (2007) p.1783
3. J. Haschke, L. Jogschies, D. Amkreutz, L. Korte, B. Rech, Solar Energy Materials & Solar Cells, 115 (2013) 7.
4. S. Varlamov, J. Dore, R. Evans, D. Ong, B. Eggleston, O. Kunz, U. Schubert, T. Young, J. Huang, T. Soderstrom, K. Omaki, K. Kim, A. Teal, M. Jung, J. Yun, Z. M. Pakhuruddin, R. Egan, M. A. Solar Energy Materials & Solar Cells (2013).
5. M. Lee, S. Moon, M. Hatano, C.P. Grigoropoulos, C. P. *Appl. Phys. Mater. Sci. Process.* **73,** 317–322 (2001).
6. J.-M. Shieh *et al.*, *Appl Phys Lett* **85,** 1232–1234 (2004).
7. B. K. Nayak, *et al., Appl. Phys. A* **80,** 1077–1080 (2005).
8. Y. J. Ma, *et al., Opt. Lett.* **36**, 3431-3433 (2011).
9. A. W. R. Leitch, J. Weber, V. Alex, Material, Science and Engineering B58, 6-12 (1999).

Mater. Res. Soc. Symp. Proc. Vol. 1666 © 2014 Materials Research Society
DOI: 10.1557/opl.2014.722

Crystallization of Amorphous Silicon and Dopant Activation using Xenon Flash-Lamp Annealing (FLA)

T. Mudgal[1], C. Reepmeyer[1], R. G. Manley[2], D. Cormier[3] and K.D. Hirschman[1]

[1]Electrical & Microelectronic Engineering Department, Rochester Institute of Technology, Rochester, New York, 14623, USA

[2]Corning Incorporated, Science and Technology, Corning, New York, 14870, USA

[3]Industrial & Systems Engineering Department
Rochester Institute of Technology, Rochester, New York, 14623, USA

ABSTRACT

Flash-lamp annealing (FLA) has been investigated for the crystallization of a 60 nm amorphous silicon (a-Si) layer deposited by PECVD on display glass. Input factors to the FLA system included lamp intensity and pulse duration. Conditions required for crystallization included use of a 100 nm SiO_2 capping layer, and substrate heating resulting in a surface temperature ~ 460 °C. An irradiance threshold of ~ 20 kW/cm^2 was established, with successful crystallization achieved at a radiant exposure of 5 J/cm^2, as verified using variable angle spectroscopic ellipsometry (VASE) and Raman spectroscopy. Nickel-enhanced crystallization (NEC) using FLA was also investigated, with results suggesting an increase in crystalline volume. Different combinations of furnace annealing and FLA were studied for crystallization and activation of samples implanted with boron and phosphorus. Boron activation demonstrated a favorable response to FLA, achieving a resistivity $\rho < 0.01$ Ω•cm. Phosphorus activation by FLA resulted in a resistivity $\rho \sim 0.03$ Ω•cm.

INTRODUCTION

Recent advancements in large-format display devices have pushed hydrogenated amorphous silicon (a-Si:H) to its performance limit. Low-temperature polycrystalline silicon (LTPS) and metal-oxide based thin-film transistors (TFTs) are candidates for replacement of a-Si:H for high-performance applications. Excimer laser annealing (ELA) based LTPS is in manufacturing, however due to its high-cost and increased process complexity it has been commercialized only for small format displays. The applicability of ELA for Gen10 glass is questionable, and the flat-panel display industry is searching for alternative LTPS strategies which are cost-effective and easily scalable to large glass panel production.

FLA is a technique which anneals the material using a series of short but intense bursts of broad spectrum light from xenon flash lamps, and can readily be extended to accommodate arbitrarily large substrates. While FLA has been shown to crystallize thin a-Si films on glass [1, 2], the reports on fabricated TFTs have been very limited [3]. This work investigates the ability to crystallize a-Si using FLA, and to activate implanted boron and phosphorus as dopants for TFT fabrication. The influence of adding a small amount of nickel to the silicon layer is also investigated as a means to enhance crystallization using the FLA process [4].

91

EXPERIMENT

A 60 nm a-Si:H layer was deposited on 150 mm diameter Corning EAGLE XG® display glass wafers using PECVD. The deposition was performed using SiH_4 and H_2 at 400 °C, 1 Torr pressure and 30 W RF power. The samples were dehydrogenated at 450 °C for 2 h. A 100 nm SiO_2 capping layer/screening oxide was deposited over the a-Si film using PECVD with TEOS precursor at 380 °C. These samples were then subjected to a variety of FLA treatment conditions, with substrate heating providing a steady-state surface temperature of ~ 460 °C as measured using an infrared thermometer. The FLA system used in this work is a NovaCentrix PulseForge 3300 configured with two Xe lamps, with a 75 mm × 150 mm exposure window. A Woollam VASE system and HORIBA Jobin Yvon Raman spectroscopy system was used for material characterization.

For NEC samples, ~6 nm of Ni was RF sputtered directly on the a-Si layer. This Ni is then etched in a mixture of phosphoric, acetic and nitric acid at 50 °C. A thin-layer of nickel silicide ($NiSi_x$) remained after etching, as indicated by a low sheet-resistance measurement (Rs ~ 750 $\Omega/\square$) following the chemical etch. While a capping oxide layer was not used for NEC samples, substrate heating was applied during FLA exposure.

For dopant activation experiments, a high fluence ($\phi = 4 \times 10^{15}$ cm^{-2}) of boron and phosphorus at energies of 35 keV and 75 keV, respectively, were implanted through the 100 nm oxide layer into the silicon film using a Varian 350D ion implanter. Some of the samples received FLA or furnace annealing treatments prior to implantation, as indicated in table 1. Nickel-exposed samples were not included. Dopant activation was quantified by the measured four-point probe sheet resistance.

DISCUSSION

The FLA system used in this study has built-in constraints on the allowable combination of lamp voltage, pulse duration and pulse frequency settings which are based on limitations on charge supply and lamp intensity. Initial trials established the importance of substrate heating and the oxide capping layer. An experiment was designed to find optimal settings for the crystallization of a-Si films.

FLA process optimization

Several combinations of pulse duration and lamp intensity were explored. A scatter plot of treatment combinations is shown figure 1, where each treatment combination has a pulse duration set to the system limit based on the lamp voltage (intensity) setting. The pulse at the highest intensity (38 kW/cm^2) was limited in duration, and did not provide enough energy for crystallization. The longest pulse duration resulted in the highest radiant exposure (8.75 J/cm^2) however the intensity was too low to deliver the energy in a short enough time for crystallization to occur. Successful crystallization requires a high energy to be delivered in a short time period, resulting in maximum silicon temperature with minimal heat transfer to the substrate. This was found to be at an intensity setting $I = 19$ kW/cm^2 and pulse duration $t = 270$ μs, providing a radiant exposure $E \sim 5$ J/cm^2.

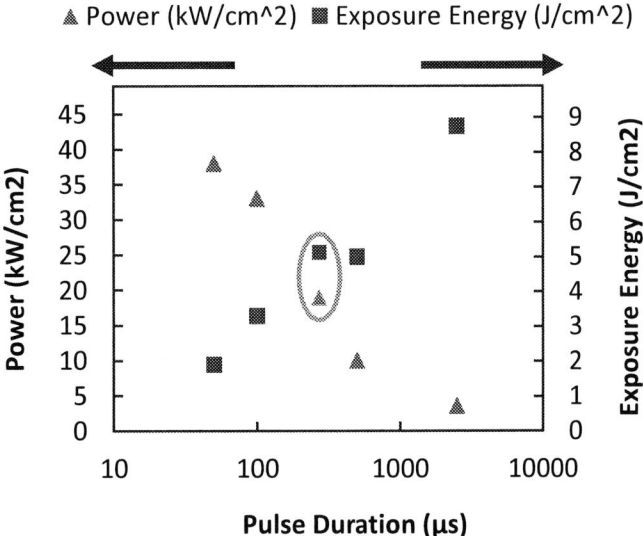

Figure 1. Scatter plot of FLA treatment combinations. Each treatment combination has a unique pulse duration which was limited by the system constraints. Each treatment has a marker for power (triangle) and a marker for exposure energy (square). The circled treatment identifies near-optimum settings for crystallization, with a radiant exposure ~ 5 J/cm^2 at an irradiance threshold ~ 20 kW/cm^2.

FLA crystallization

The prepared samples were FLA exposed using the optimized settings established. Samples were then optically characterized by VASE analysis and reflection spectra. Using a Tauc-Lorentz oscillator model, the refractive index (n) and extinction coefficient (k) for a-Si and FLA-treated silicon layer were developed and are shown with a crystalline silicon reference in figure 2. These results confirm a transformation to polycrystalline material with only a single pulse exposure, as does the Raman spectrum for sample c shown in figure 3 which demonstrates a characteristic peak at 517 cm^{-1}, near that of crystalline silicon.

The SiO$_2$ capping layer may serve as an anti-reflective layer for FLA exposure, or as a thermal barrier to avoid substantial surface heat loss during the FLA process, or both. It is required to realize crystallization in a single pulse on FLA samples processed without any nickel introduction, as shown by the difference between samples b & c in figure 3. The nickel-enhanced NEC-FLA sample was FLA exposed without any SiO$_2$ capping layer. After a single pulse at optimized settings the film became visually transparent, whereas samples processed without surface nickel and without a SiO$_2$ capping layer required between 5-10 pulses to see visible evidence of crystallization. Note that for multiple pulse exposures the repetition rate (frequency) used at the optimum intensity & time settings was 2 Hz, limited by the system constraints.

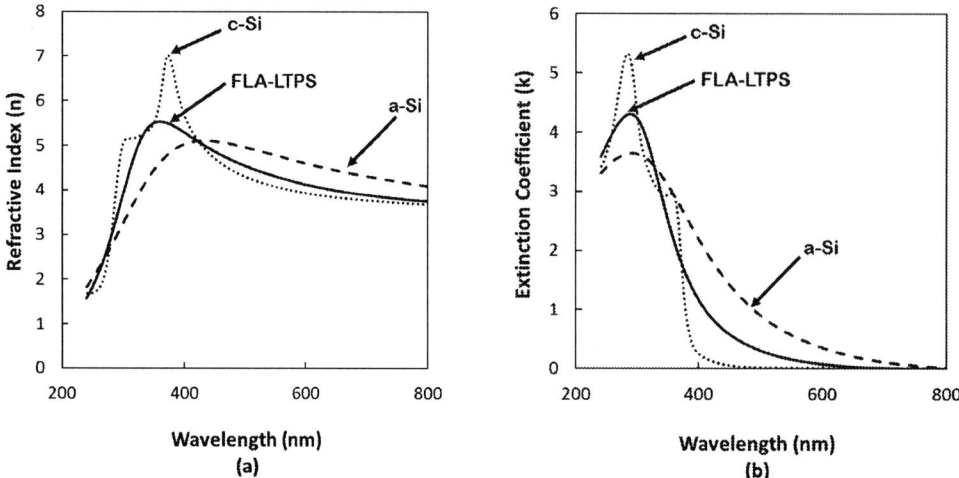

Figure 2. (a) Refractive index and (b) extinction coefficient of c-Si (dotted line), FLA (solid line) and a-Si (dashed line) over λ range from 240 nm to 800 nm. Both optical constants of the FLA sample confirm a conversion to polycrystalline silicon. Absorption decreases at wavelengths above 450 nm, which was obvious in a visual comparison with a-Si.

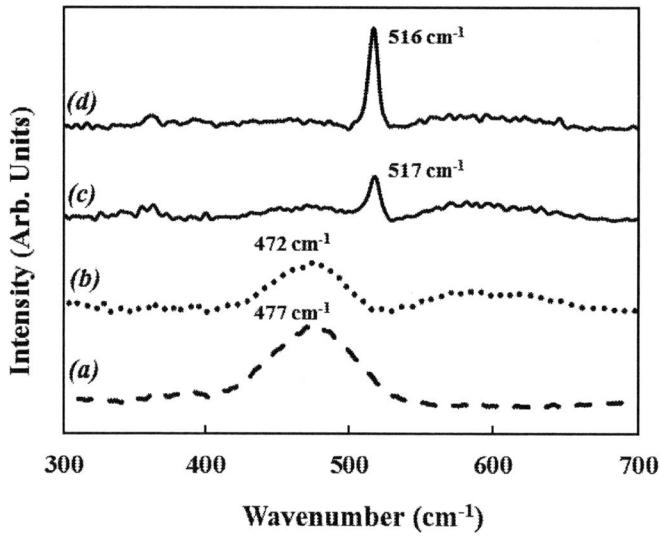

Figure 3. Raman spectra of samples following a single-pulse FLA exposure (except sample *a*) with peak positions indicated. Samples include a-Si (*a*), FLA processed samples without (*b*) and with (*c*) a 100 nm SiO_2 capping layer, and a NEC-FLA processed sample without an oxide capping layer (*d*). The comparison of samples *b* & *c* demonstrates the importance of the SiO_2 capping layer if Ni is not introduced. The comparison of samples *c* & *d* indicates that the introduction of Ni results in enhanced crystallization, supported by the reduction in FWHM (9.1 cm^{-1} versus 7.2 cm^{-1}). The peak positions in samples *c* & *d* are slightly offset from crystalline Si (520 cm^{-1}), which is likely due to residual stress in samples after FLA.

The NEC-FLA treatment was difficult to evaluate using VASE, with a significant discrepancy between the optical model and measured data. This is likely due to a remaining surface silicide layer, however adding more complexity to the optical model was not pursued. A comparison of NEC-FLA processed film with FLA-only samples, processed both with and without a SiO_2 capping layer, and a-Si is on the Raman spectra shown in figure 3. The peak intensity of sample *d* is slightly shifted and more pronounced than sample *c*, with a reduction in FWHM which supports an increase in crystalline phase. Multiple pulses (10) on the NEC-FLA sample resulted in further reduction in FWHM to 5.6 cm^{-1} (not shown).

Dopant activation

Investigation on the activation behavior of implanted boron and phosphorus for LTPS using solid-phase crystallization (SPC) by furnace annealing was reported previously [5], with select results included here for comparison. Table 1 shows the conditions for a number of samples processed using combinations of furnace annealing and FLA. Key findings from the previous work that are important for this discussion are represented by treatments (T#) B1, B2, P1 and P2. All furnace anneals, whether they were pre-implant or post-implant, were done at 630 °C for 12 h. All FLA exposures were at the optimized settings, with count number (#) specified. Sheet resistance (R_S) results indicate the level of dopant activation, with "X" indicating that activation was either minimal or unsuccessful. Note that the typical R_S range within samples was ~ ±10%, so the values listed are approximate.

Table 1. Dopant activation treatments for comparison.

T#	Pre-implant anneal	Implant	Post-implant anneal	Rs (Ω/□)
B1	None	Boron	Furnace	900 *
B2	Furnace	Boron	Furnace	900 *
B3	None	Boron	FLA (1#, 20#)	1400, 1200
B4	None	Boron	FLA + Furnace	20k (deactivation)
B5	FLA (10#)	Boron	FLA (1#, 10#, 20#)	X
B6	Furnace	Boron	FLA (10#, 20#)	600, 500
P1	None	Phosphorus	Furnace	X *
P2	Furnace	Phosphorus	Furnace	1500 *
P3	None	Phosphorus	FLA (5#, 10#, 20#)	25k, 10.7k, 4600
P4	FLA (10#)	Phosphorus	FLA (1#, 10#, 20#)	X
P5	FLA (1#)	Phosphorus	FLA (1#)	4400
P6	Furnace	Phosphorus	FLA (20#)	2300

* furnace anneal results reported previously [5]

Boron-implanted treatments B1 and B2 demonstrated that the degree of boron activation using furnace annealing did not depend on whether or not furnace SPC was done prior to the implant [5]. Treatment B3 demonstrates that FLA can be effective in activating boron with R_S decreasing between 1# and 20#, however the sheet resistance is not as low as B1. Treatment B4 was the same as B3 (20#) followed by a furnace anneal, and demonstrated a marked increase in R_S due to boron deactivation. Treatment B5 with a pre-implant FLA treatment was not successful in demonstrating boron activation, which establishes a fundamental difference in the behavior of the FLA-crystallized material compared to the furnace SPC process. Treatment B6 was the same as B2 except the post-implant anneal was done using FLA instead of a furnace anneal. The lower R_S value indicates an improvement in activation that continues using multiple (20#) FLA exposures. Treatments B3 and B6 have acceptable levels of activation for TFT applications, with an improvement observed if furnace SPC is done prior to boron implantation.

Phosphorus-implanted treatments P1 and P2 demonstrated that furnace SPC was required to realize furnace activation using a furnace anneal [5]. Treatment P3 demonstrates that FLA can activate phosphorus without a pre-implant anneal, unlike the furnace activation process, however a high pulse-count was required and the activation level was still limited. Treatment P4 demonstrates that FLA with multiple counts performed prior to phosphorus implantation does not support FLA activation of phosphorus. However treatment P5 which only received a single FLA pulse pre-implant and post-implant achieved a similar level of activation as P3 with 20# FLA exposure. Treatment P6 was the same as P2 except the post-implant anneal was done using FLA (20#) instead of a furnace anneal. The furnace SPC prior to phosphorus implantation enhances phosphorus activation, as in the boron case. However unlike boron, the FLA process did not achieve the same level of phosphorus activation as a furnace anneal.

CONCLUSIONS

This study investigated the ability of FLA to crystallize thin-film a-Si deposited on display glass, and activate implanted boron and phosphorus. Initial process runs established the importance of substrate heating and a SiO_2 capping layer. Optimized process settings were found at an intensity threshold ~ 20 kW/cm^2 delivering a radiant exposure ~ 5 J/cm^2 within a pulse width ~ 250 μs. The crystallization process was validated using VASE analysis and Raman spectroscopy. The introduction of Ni was found to enhance the FLA crystallization process, resulting in a decrease in the Raman peak FWHM. Implanted boron and phosphorus were both electrically activated using FLA, however only boron results appear to be suitable for TFT applications. Additional material characterization is needed to reveal the differences between furnace SPC and FLA crystallized films, and investigate certain activation behavior anomalies such as treatments B4 and P5 in table 1.

ACKNOWLEDGMENTS

The authors would like to acknowledge the support of the technical staff at the Semiconductor & Microsystems Fabrication Laboratory at RIT, and the technical staff at Corning that provided analytical services. Financial support has been provided by Corning Incorporated and NYSTAR, through the New York State Center for Advanced Technology.

REFERENCES

1. K. Ohdaira, T. Fujiwara, Y. Endo, S. Nishizaki and H. Matsumura, *J. Appl. Phys.*, **106**, 044907 (2009).
2. K. Ohdaira and H. Matsumura, *Journal of Crystal Growth*, **362**, 149 (2013).
3. S. Saxena, D. C. Kim, J. H. Park and J. Jang, *IEEE Electron Device Letters*, **31**, 1242 (2010).
4. Z. H. Jin, G. A. Bhat, M. Yeung, H. S. Kwok and M. Wong, *J. Appl. Phys.*, **84**, 194 (1998).
5. Q. Li, T. Mudgal, P. M. Meller, S. Slavin, R. G. Manley and K. D. Hirschman, *MRS Online Proceedings Library*, **1426**, 281 (2012).

Mater. Res. Soc. Symp. Proc. Vol. 1666 © 2014 Materials Research Society
DOI: 10.1557/opl.2014.797

Defects and Doping in Nanocrystalline Silicon-Germanium Devices

Siva Konduri, Watson Mulder and Vikram L. Dalal
Department of Electrical and Computer Engineering, Iowa State University, Ames, Iowa

ABSTRACT

Nanocrystalline Silicon-Germanium (Si,Ge) is a potentially useful material for photovoltaic devices and photo-detectors. Its bandgap can be controlled across the entire bandgap region from that of Si to that of Ge by changing the alloy composition during growth. In this work, we study the fabrication and electronic properties of nanocrystalline devices grown using PECVD techniques. We discovered that upon adding Ge to Si during growth, the intrinsic layer changes from n-type to p-type. We can change it back to n-type by using ppm levels of phosphorus, and make reasonable quality devices when phosphine gas was added to the deposition mix. We also measured the defect density spectrum using capacitance frequency techniques, and find that defect density decreases systematically as more phosphine is added to the gas phase. We also find that the ratio of Germanium to Silicon in the solid phase is higher than the ratio in the gas phase.

INTRODUCTION

Nanocrystalline Si (nc-Si) is an important and useful material for solar cells and photo-detectors. Significant progress has been made in the solar conversion of efficiency of nc-Si with efficiencies exceeding 10% in single junction cells, and exceeding 15% in multiple junction cells[1-5]. However, there has been little progress in increasing the efficiencies of nanocrystalline (Si,Ge) alloys[6-10], which can potentially achieve high quantum efficiencies in the infrared range out to 0.68 eV i.e. the bandgap of crystalline Ge. It is usually found that as one adds Ge to Si, the device performance becomes worse, and it has been speculated that this decrease in performance is due to additional Ge-associated defects in the material [10, 11]. In this paper, we show that addition of Ge to Si leads to significant increases in defect density, and may even make the intrinsic n layer material to p-type. We also show that addition of ppm levels of phosphorus, using phosphine, to the intrinsic layer reduces the defect density and makes the material n type again, and allows one to achieve reasonable devices with quantum efficiencies extending further into the infrared regions.

EXPERIMENTS

The materials and devices were fabricated using PECVD techniques using a VHF plasma at ~47 MHz. The substrate temperature was in the range of ~250 °C and pressure was ~ 100mT. The precursor gases were hydrogen, silane and 10% germane diluted in hydrogen. Varying levels of Ge:Si could be produced by varying the ratio of germane to silane flow ratio. Ppm levels of dopants such as phosphorus and boron could be introduced in the base (intrinsic) layer using dopants such as phosphine and diborane diluted in hydrogen. The devices were standard p+in+ devices, deposited on pre-cleaned stainless steel substrates with the i layer generally being n type

97

for nc-Si. ITO dots are deposited as top contacts and contact area is $0.12 cm^2$. SiGe is used as p-layer to avoid kinks in bandgap and a smooth transition at p-i interface and Ge concentration in p-layer is around 20%. The basic device structure of nc-SiGe:H solar cell is shown in Fig. 1. Thickness of i-layer is about 0.70 μm.

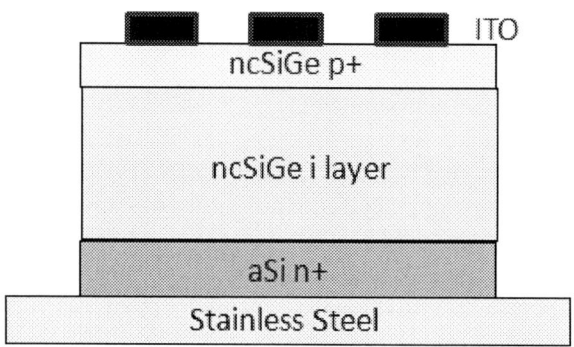

Figure 1. Typical device structure of nc-SiGe:H solar cell

For measuring defect densities, we used the capacitance-frequency techniques described previously [12]. Device quantum efficiency (QE) was also measured using standard techniques. The ratio of Ge:Si in the solid phase was determined using energy dispersive x-ray spectroscopy.

RESULTS

In Fig. 2, we show the ratio of Ge:Si in the solid phase for various germane:silane flow rates. It is clear from this figure that much more Ge is incorporated in the solid phase as than is present in the gas phase.

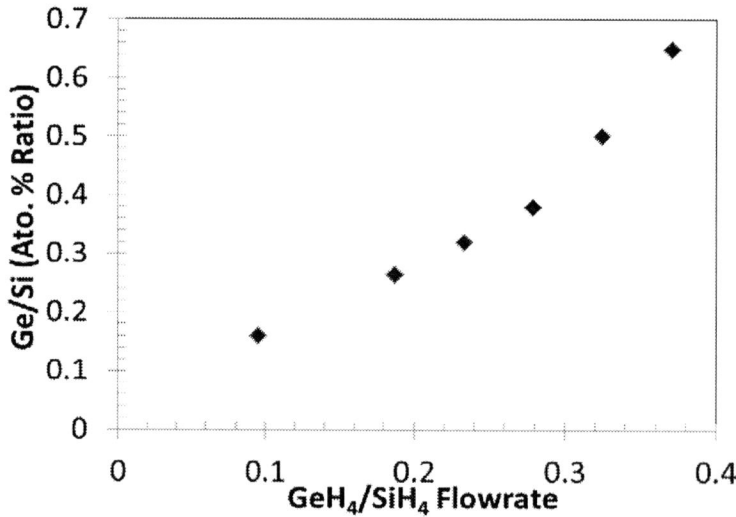

Figure 2. Germanium content for different gas flows in nc-SiGe:H devices

In Fig. 3a, we show the illuminated I-V curves for samples prepared using varying germane:silane ratios. A surprising result is that the short-circuit current actually decreases as the germane content in the gas phase increases. This is in spite of the fact that the germane:silane ratio in the solid phase is increasing, thereby implying a smaller bandgap material for higher germane flows. The explanation for this behavior is provided by studying quantum efficiency data, shown in Fig. 3b. It is clear from the QE data that there is a drastic reduction in QE for short wavelengths when significant amounts of germane is added to the input gas mixture. This can be ascribed to poor collection of electrons being generated near the p-n interface by light of shorter wavelengths, i.e. to a drastic reduction in electron diffusion length. This could be a result of the n-type intrinsic layer changing to a p-type, presumably due to Ge defects absorbing electrons from the oxygen induced donor states [13]. Indeed, there appears to be a significant increase in defect density when Ge content increases. See Fig. 4 for data on defect density vs. energy for varying German:Silane flow rates.

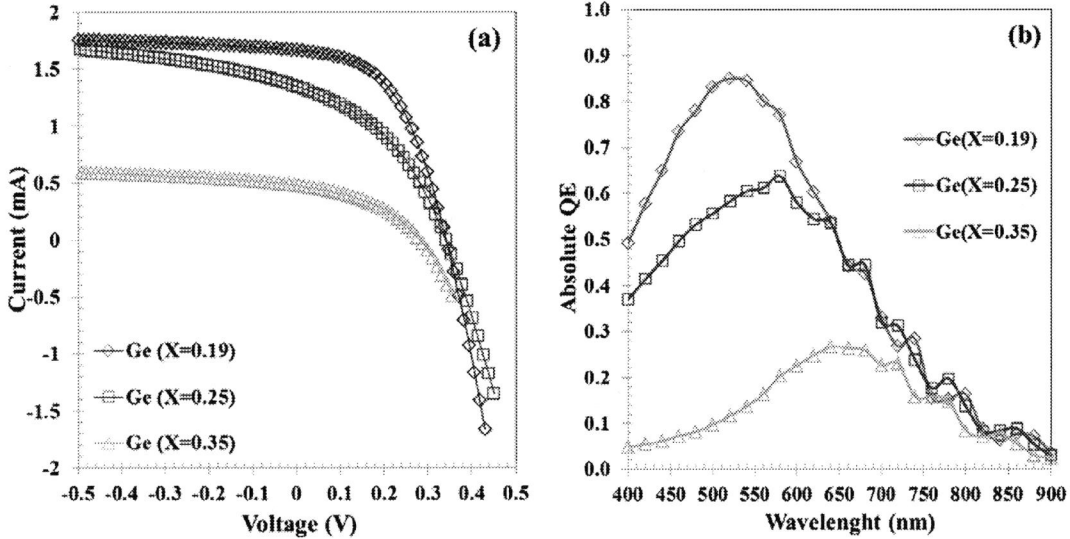

Figure 3. (a) IV curves and (b) Absolute QE vs. Wavelength for nc-SiGe:H solar cells with varying Ge content

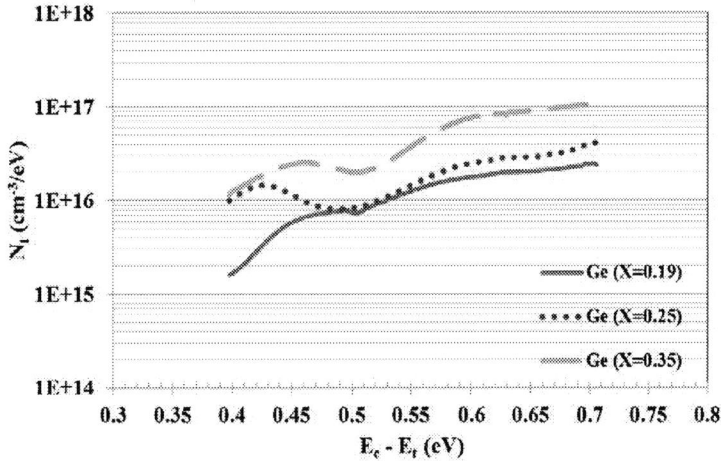

Figure 4. Calculated defect density vs. energy below the conduction band for varying Ge content

To overcome the deleterious effects of additional Ge on both defect density and QE, we tried to compensate for the apparent p-type doping by adding ppm levels of PH_3 to the gas mixture. The results for I-V curve are shown in Fig. 5a for various values of PH_3 for a fixed germane:silane ratio of in the gas phase. Clearly, adding PH_3 significantly improves the short-circuit current in the solar cell. The corresponding QE curves are shown in Fig. 5b, and they clearly show an improvement in QE at short wavelengths, compared to the QE curves shown in Fig. 3, thus showing that the poor electron collection efficiency for photons which are absorbed near the p-intrinsic layer interface has been overcome. The corresponding change in total defect density as a function of PH_3 flow is shown in Fig.6, thus confirming that addition of ppm levels of PH_3 serves to reduce the defect density in nanocrystalline (Si,Ge) intrinsic layers, probably by compensating the defects caused by Ge addition.

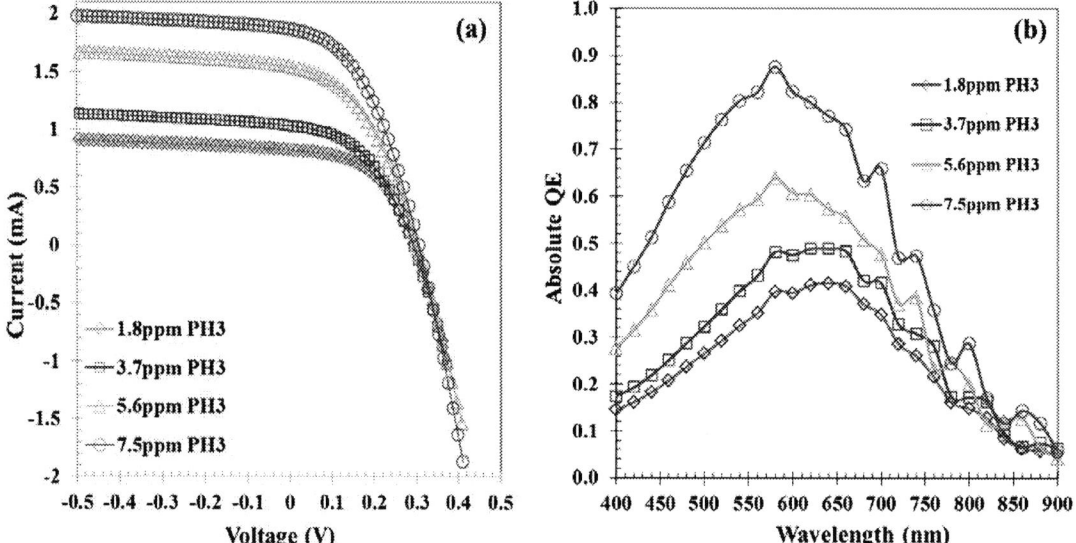

Figure 5. (a) IV curves and (b) Absolute QE vs. Wavelength for nc-SiGe:H solar cells with (X_{Ge}~0.35) for varying PH_3 flow rates

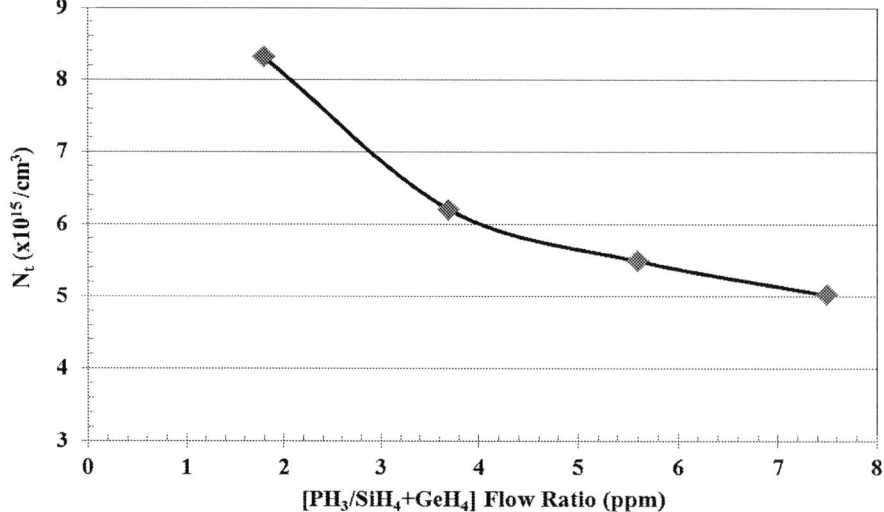

Figure 6. Total defect density as a function of PH_3 flow rate

I-V curves for the devices prepared using varying ratios of germane:silane flow along with optimum ppm levels of PH_3 are plotted in Fig.7a, which clearly shows that increasing Ge content leads to higher short-circuit currents but approximately the same open-circuit voltage. In general, V_{oc} depends on quality of i-layer, type of p-layer and transition layers. Further experiments are being conducted to examine the relationship of Voc with Ge content. The corresponding QE data in Fig. 7b shows that increasing Ge content shifts the curves to longer wavelengths, as expected.

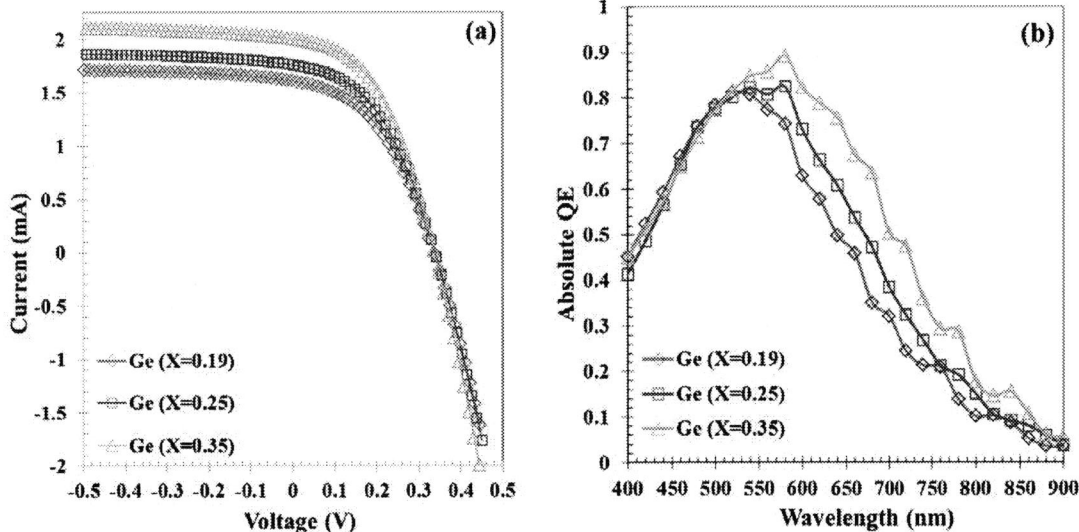

Figure 7. (a) IV curves and (b) Absolute QE vs. Wavelength for nc-SiGe:H solar cells with varying Ge content with optimum PH_3 flow (i-layer Thickness ~ 0.7μm)

CONCLUSIONS

In conclusion, we have shown that the defect density in the intrinsic layer of nanocrystalline p-i-n solar cells material increases significantly as Ge content increases. It appears that adding Ge changes the doping type of the intrinsic layer form n-type to p-type. This shift leads to a drastic reduction in quantum efficiency for short wavelength photons at higher Ge concentrations. The defect density can be reduced, and the p-type behavior changed back to n-type by adding ppm levels of PH_3 to the gas phase. The results show that when appropriate compensation by Phosphorus has been achieved, the quantum efficiency for solar cells extends further out towards longer wavelengths as the Ge content of the solar cell intrinsic layer increases, as expected, resulting in larger currents. This is an important result which shows that nanocrystalline (Si,Ge) cells can be used to improve the efficiency of tandem junction solar cells, as also the infrared response when these materials are used as photo-detectors.

ACKNOWLEDGMENTS

It is a pleasure to acknowledge the technical help of Andrew Gulstad and Max Noack. This work was partially supported by a grant from NSF.

REFERENCES

1. J. Meier, R. Flückiger, H. Keppner and A. Shah, Applied Physics Letters, **65**, 860-862 (1994)
2. A. Shah, J. Meier, E. Vallat-Sauvain, C. Droz, U. Kroll, N. Wyrsch, J. Guillet and U. Graf, Thin Solid Films, **403–404**, 179-187, (2002)
3. A. Shah, J. Meier, E. Vallat-Sauvain, N. Wyrsch, U. Kroll, C. Droz and U. Graf, Solar Energy Materials and Solar Cells, **78**, 469-491, (2003)
4. V. Dalal, J. Leib, K. Muthukrisnan, D. Stieler, and M. Noack, 2005 IEEE Photovoltaic Specialists Conference, 1448- 1451, (2005)
5. G. Yue, B. Yan, L. Sivec, T. Su, Y. Zhou, J. Yang and S. Guha, 2012 MRS Proceedings, **1426**, 33-38, (2012)
6. M. Isomura, K. Nakahata, M. Shima, S. Taira, K. Wakisaka, M. Tanaka and S. Kiyama, Solar Energy Materials and Solar Cells, **74** (1–4), 519-524 (2002)
7. J. K.Rath, F.D. Tichelaar and R. E.I Schropp, Solar energy materials and solar cells, **74,** 553-560 (2002)
8. S. Saripalli and V. Dalal, EIT 2008 IEEE International Conference Proceedings, 414-418, (2008)
9. T. Matsui, M. Kondo, K. Ogata, T. Ozawa and M. Isomura, Applied Physics Letters, **89**, 142115, (2006)
10. T. Matsui, K. Ogata, C.W. Chang, M. Isomura and M. Kondo, Journal of Non-Crystalline Solids, **354** (19–25), 2468-2471, (2008)
11. T. Matsui, C.W. Chang, T. Takada, M. Isomura, H. Fujiwara and M. Kondo, Solar Energy Materials and Solar Cells, **93**, 1100-1102, (2009)
12. K. Shantan, K. Siva and V. Dalal, Applied Physics Letters, **103**, 093506 (2013)
13. T. Matsui, C.W. Chang, K. Mizuno, Y. Takeuchi and M. Kondo, Japanese Journal of Applied Physics, **51,** 091302, (2012)

Mater. Res. Soc. Symp. Proc. Vol. 1666 © 2014 Materials Research Society
DOI: 10.1557/opl.2014.911

Tunable and Wireless Photoimpedance Light Sensor

Tanuj Saxena[1], Sergey Rumyantsev[1], Partha Dutta[1] and Michael Shur[1,2]

[1] Department of E.C.S.E, Rensselaer Polytechnic Institute, Troy, NY 12180, USA

[2] Department of Physics, Applied Physics and Astronomy, Rensselaer Polytechnic Institute, Troy
NY 12180, USA

ABSTRACT

We report on the effects of the frequency dispersion in light sensitive materials used in photoimpedance wireless sensors. An example of such a sensor is a gated semiconductor connecting two or more fixed capacitances. The impedance of the device under illumination is changed by the change in the photoresistance of the semiconductor layer and the change in the gate-semiconductor capacitance. We report on the design and simulation of the frequency dispersion of the impedance of this device in silicon and discuss the physics and device performance. We also evaluate the dynamic range and sensitivity of the wireless photoimpedance sensors and show their advantages for wireless sensing applications compared to more conventional light sensors.

INTRODUCTION

The emergence of LED based intelligently controlled lighting systems stimulated growing interest in versatile wireless sensors capable of color discrimination and having a wide dynamic range and high sensitivity. In this paper, we report on photocapacitive sensors utilizing the frequency dispersion for expanding the dynamic range. Such sensors can easily be integrated into wireless RF circuits. Conventional photocapacitive devices like Schottky diodes [1, 2] work only on change in trap occupancy in the space charge region under illumination, which is usually small. Demonstrated MOS photocapacitors [3-5] use the modulation of inversion layer under illumination. We propose a novel design, which utilizes the physical mechanisms changing both *active and imaginary* device impedance (i.e. resistance and capacitance). This sensor design consists of two or more fixed capacitances connected by the light and frequency sensitive elements. We have demonstrated such a concept using a cadmium sulfide device [6]. In this report, we present the simulation results for a related but different structure implemented in silicon.

DEVICE DESIGN

Fig. 1. Shows the device design that consists of fixed capacitors (geometric capacitors) monolithically integrated with photosensitive elements - a metal-oxide-semiconductor (MOS) capacitance and semiconductor photoresistance. The bottom plates of the geometric capacitances sit directly on the semiconductor substrate and are separated from the top plates by a dielectric film. The top plates of the geometric capacitances and the gate of the MOS capacitor structure are tied together to form the gate terminal of the device. The second terminal of the device is the substrate of the MOS structure. A narrow metal line directly under the gate serves as the substrate contact. Illumination leads to changes in the MOS capacitance and in the reduction of the resistance between the substrate contact and the bottom plates of the geometric capacitances. Thus the overall capacitance (or complex impedance) between the gate and substrate terminals is changed by change in the MOS capacitance and its coupling with the geometric capacitances.

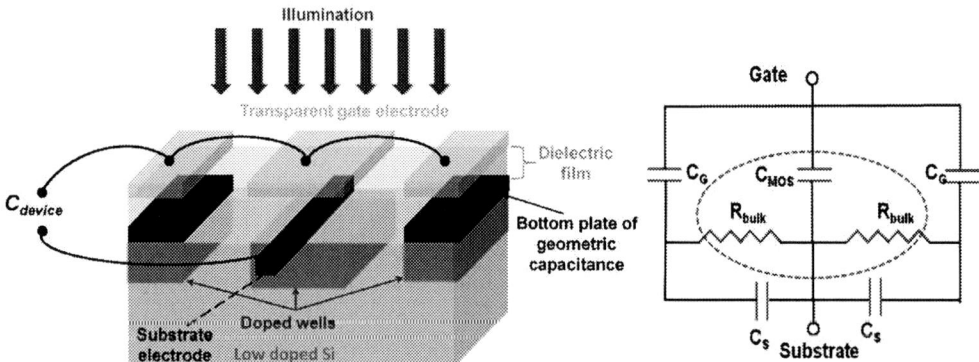

Figure 1. Schematic of the device structure and a simple lumped model of the device. C_G - geometric capacitance, C_{MOS} - MOS capacitance, R_{bulk} - semiconductor photoresistance and C_S - semiconductor dielectric capacitance.

SIMULATION DETAILS

Sentaurus TCAD was employed to design and simulate the device. The photoresistive connection between the different capacitors could be created by either having a p-n junction or a low-doped semiconductor layer. Low doping ensured a large dark resistance improving the photosensitivity of the semiconductor layer in the range of intensities chosen. The simulated structure had a low doped n-type silicon substrate (5×10^{11} cm^{-3}) and three wells with a higher concentration (5×10^{14} cm^{-3}) (Fig. 1). One of these wells acted as the substrate of the MOS capacitance, while the others facilitated a low contact resistance to the bottom plates of the geometric capacitors. A 200 nm thick SiO$_2$ was used as the dielectric. The gate length for the MOS capacitance was 20 µm and the gate (top plate) length for each geometric capacitance was 10 µm. The bottom plates of the geometric capacitances and the substrate contact were from aluminum (100 nm thick). The substrate contact was a narrow line, 2 µm wide, while the bottom plates of the geometric structures were 10 µm wide each. Aluminum was employed also as the gate metal for the MOS capacitance and the top plates of the geometric capacitances. However, in a real device, a transparent metal like ITO or a grid-gate structure should be employed to let light enter the photosensitive regions of the device.

DISCUSSION

Small signal ac simulations were carried out to obtain the impedance dispersion of the device at different levels of illumination. Monochromatic light of 400 nm was used as the illumination source within the simulation framework. The capacitance (C_{device}) was simulated between the gate and the substrate. Fig. 2 shows the small signal C-V characteristics for the device capacitance at the ac frequency of 10 kHz under different illumination levels. Fig. 3 shows the hole and electron concentration profiles under dark and illumination with the gate bias of -5V. At this bias, the MOS capacitance is under inversion, as can be seen from the concentration profile of holes. The electrons, on the other hand, are depleted and the region under the substrate contact also shows depletion. Under illumination, the generated holes contribute to changing the semiconductor bulk resistance as well as to the modulation of the MOS inversion channel. The generated electrons predominantly contribute to the change in the semiconductor bulk resistance. Under illumination, the MOS capacitance value increases towards the oxide capacitance value and with the reduction in bulk semiconductor resistance, the coupling of the MOS capacitance with the geometric capacitances becomes stronger, leading to a further increase in the device capacitance. Hence, this design leads to a larger capacitance swing as compared to the MOS capacitor alone. It also allows tailoring the floor and ceiling of the device capacitance by changing the geometries of MOS and geometric capacitances and their separation independently.

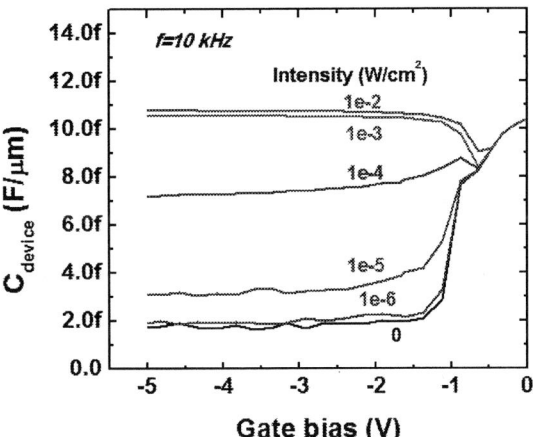

Figure 2. Variation of small signal capacitance, C_{device}, as a function of gate bias under different illumination intensities (ac frequency = 10 kHz).

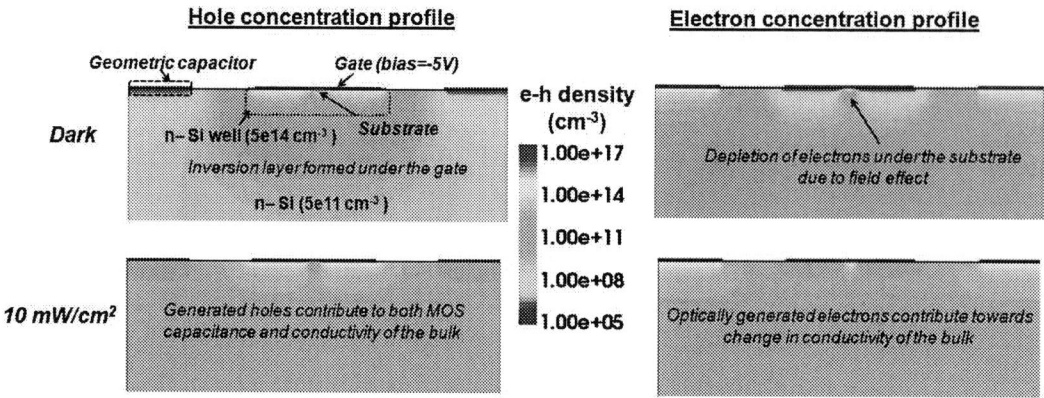

Figure 3. Hole and electron concentration profiles in dark and under illumination intensity of 10 mW/cm^2 at gate bias of -5V.

Fig. 4 shows the frequency dispersion of the device capacitance under different illumination intensities. At smaller frequencies, the device capacitance is high as the MOS capacitance is close to the oxide capacitance value and the bulk resistance of the semiconductor is less significant in determining the impedance (as the impedance from geometric capacitances becomes larger at lower frequencies). As the frequency is increased, the capacitance drops due to both a decrease in the MOS capacitance and the increasing significance of the bulk semiconductor resistance. The MOS capacitance reduces at higher frequencies because the thermal generation rate can't keep up with the high frequency signal on the gate and as a result the inversion layer can't be modulated. Under illumination, the carrier generation rate increases and in effect the illumination intensity and the ac frequency compete with each other. As illumination intensity is increased, the capacitance at every frequency tends to increase due to the effects described earlier. The interesting feature of these dispersion characteristics is that the device capacitance changes from its highest to its lowest value over a narrow band of frequencies. With increase in the illumination intensity, this band is progressively shifted to higher frequencies.

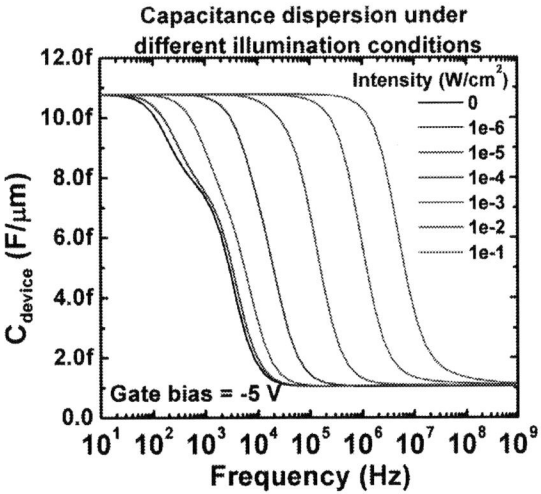

Figure 4. Frequency dispersion of device capacitance (C_{device}) under different illumination intensities.

This picture can be transformed to that in Fig. 5 where the variation of capacitance with intensity is shown for different ac frequencies. It can be observed from this figure that at any given frequency, the device capacitance is sensitive to a narrow range of intensities, but this range is different for different frequencies. At low frequencies, the sensitivity is high at relatively lower intensities and shifts to higher intensities as the frequency is increased. Thus by tuning the frequency of measurement, the sensitivity of the sensor could be tuned. Moreover, by sweeping the frequency over a large range, this sensor can exhibit a wide dynamic range without sacrificing sensitivity in individual segments. Thus, by using the frequency dispersion, the sensor becomes tunable and more versatile than its dc counterparts.

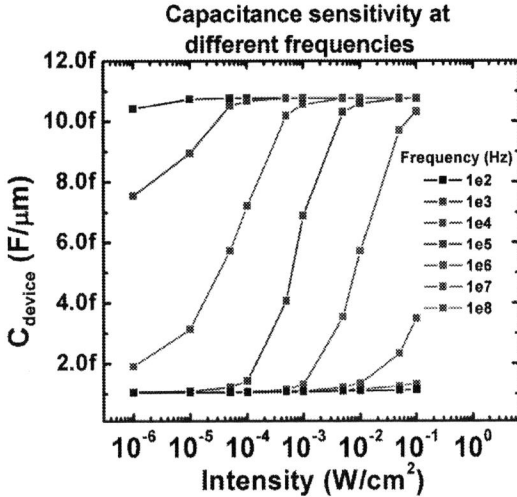

Figure 5. Device capacitance (C_{device}) as a function of illumination intensity at different ac frequencies (gate bias = -5V).

From Fig. 2, it can also be observed that the small signal capacitance becomes virtually independent of bias at bias voltages less than -2 V. Thus, with appropriate biasing, the large signal capacitance would be the same as the small signal capacitance and would follow the same trend with frequency and illumination. This makes the device quite versatile and it can be employed as a light sensitive element in a variety of circuits. A straightforward implementation of the device would be in the tank circuit of an oscillator. As the capacitance is changed, the frequency of oscillation would change and this change can be detected wirelessly.
A control circuit can further be added to allow different values of "dark frequency" of oscillation and hence enable sensing of a wide dynamic range of intensities.

CONCLUSIONS

We presented the simulation results of a versatile photoimpedance sensor whose frequency dispersion allows for tunability of sensitivity and dynamic range. Low illumination intensities with high accuracy can be measured at low measurement frequencies, whereas higher measurement frequencies should be employed to sense higher intensities. This device was designed with low doped silicon but similar concepts would apply if a p-n junction or an SOI substrate is employed for high dark resistance. The concept of the device could also be extended to semiconductors having localized states in the gap, such as a:Si-H, where the position of the quasi-Fermi level in the energy gap determines the characteristic response time.

ACKNOWLEDGMENTS

This research work was funded by the National Science Foundation, under cooperative agreement EEC-0812056, and the New York State, under NYSTAR contract C090145.

REFERENCES

1. D. Ciplys, V. S. Chivukula, A. Sereika, R. Rimeika, M. S. Shur, X. Hu, and R. Gaska, "Wireless UV sensor based on photocapacitive effect in GaN," *Electron. Lett.*, vol. 45, no. 12, p. 653, 2009.
2. V. Chivukula, D. Ciplys, A. Sereika, M. Shur, J. Yang, and R. Gaska, "AlGaN based highly sensitive radio-frequency UV sensor," *Appl. Phys. Lett.*, vol. 96, no. 16, p. 163504, Apr. 2010.
3. D. Lopez et Al., "MOS capacitors characterization under illumination" IEEE 26th International Conference on Microelectronics, pp 583-585, 2008
4. J. Grosvalet and C. Jund, "Influence of illumination on MIS capacitances in the strong inversion region," *IEEE Trans. Electron Devices*, vol. 14, no. 11, pp. 777–780, Nov. 1967.
5. P. Chakrabarti, B. R. Abraham, A. Dhingra, A. Das, B. S. Sharan, and V. Maheshwari, "Effect of illumination on the characteristics of a proposed hetero-MIS diode," *IEEE Trans. Electron Devices*, vol. 39, no. 3, pp. 507–514, Mar. 1992.
6. T. Saxena, S. L. Rumyantsev, P. S. Dutta, and M. Shur, "CdS based novel photo-impedance light sensor," *Semicond. Sci. Technol.*, vol. 29, no. 2, p. 025002, Feb. 2014.

Mater. Res. Soc. Symp. Proc. Vol. 1666 © 2014 Materials Research Society
DOI: 10.1557/opl.2014.919

Defects in Epitaxial lift-off Thin Si Films/Wafers and Their Influence on the Solar Cell Performance

Bhushan Sopori[1], Srinivas Devayajanam[1,2], Prakash Basnyat[1,2], Helio Moutinho[1], Robert Reedy[1], Kaitlyn VanSant[1], T.S.Ravi[3], Ruiying Hao[3], Jean Vatus[3] and Somnath Nag[3]
[1]National Renewable Energy Laboratory, Golden, CO
[2]New Jersey Institute of Technology, Newark, NJ
[3]Crystal Solar Inc., San Jose, CA

ABSTRACT

In this paper, we will describe the nature of defects and impurities in thick epitaxial-Si layers and their influence on the cell efficiency. These wafers have very low average dislocation density. Stacking faults (SFs) are the main defect in epi layers. They can occur in many configurations—be isolated, intersecting, and nested. When nested, they can be accompanied by formation of coherent twins resulting in dendritic growth, with pyramids protruding out of the wafer surface. Such pyramids create large local stresses and punch out dislocations. The main mechanism of dislocation formation is through pyramids. Stacking faults degrade solar cell performance. Analyses of the solar cells have revealed that the nested SFs have a controlling effect on the solar cell performance. A well-controlled growth can minimize defect generation and produce wafers that can yield cell efficiencies close to 20%.

INTRODUCTION

To date, the Si technology has continued to dominate the solar cell/module market over other thin film photovoltaic (PV) technologies. Although, the current price of PV energy is quite low, in part because of the excess production, further cost reductions are needed to reach the SunShot goals. Because the highest-cost item in a Si solar module is the Si wafer, lowering the cost of the wafer itself can be very effective in lowering the overall cost of the module. The current high cost of the Si wafer is due to the fact that there are many process steps involved in making a standard wafer. These include growing an ingot and chopping its seed and tail ends, sawing the ingot into wafers and removing the surface damage due to sawing (kerf loss of about half of the total Si material). One approach to reduce 'per wafer cost' is to produce a wafer directly from the gas phase. Epitaxial lift off also called a DGTW (Direct Gas to Wafer) system is a technology aimed at formation of a shaped wafer directly from Si-bearing gases as an epitaxial growth.

A brief description of Epi-liftoff technology

In this method, a crystalline Si film is grown on a mono-crystalline reusable temporary-substrate that has a porous surface of suitably tailored characteristics. The epitaxially deposited film (of appropriate thickness) is separated to become a free-standing wafer for solar cell fabrication.

Alternately, cell fabrication is completed while the Si film is attached to the temporary substrate, followed by separation of the cell from the substrate. The DGTW epitaxy uses trichlorosilane as a Si bearing gas. Typically the growth is carried out around 1000 °C. Other details of the growth are available in reference [1]. Typically, the quality of the material is very high, yielding best cell efficiency more than 20%.

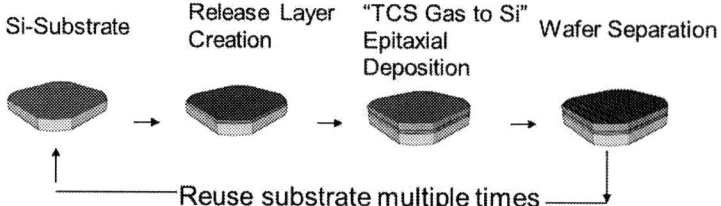

Figure 1. Epi-lift off approach in which Si wafer is directly generated by growing an epitaxial layer on a mono-crystalline substrate

However, as with any crystal growth technique, this approach also has propensity for generating certain crystal defects if the conditions of the growth are not appropriate. Likewise, impurities can also be incorporated into the growing crystal that could impact the final performance of the solar cell. Because the solar cell performance depends on the nature of defects and their interaction with impurities, it is imperative to perform detailed analyses of the defects, impurities, and the minority carrier lifetime of the wafers.

The mechanism(s) of defect formation are not yet completely understood. Although cell efficiencies comparable to mainstream silicon PV are possible, it is expected that understanding the mechanisms of the specific defects, impurities, and impurity-defect interactions can lead to their mitigation with a concomitant increase in the cell performance. Hence, we have begun a study to determine the mechanism(s) that limit the efficiency of current cells and establish approaches to overcome these limitations. In this paper, we will describe the results of a study of defects and impurities in the epitaxial Si layer and their influence on the cell efficiency.

EXPERIMENTAL PROCEDURE

Because the current wafers have low defect densities, it was helpful to include wafers from previous vintages. We collected wafers grown under a variety of conditions over a long period of time. They consisted of 150 μm and 100 μm thick wafers, both P and N type grown under several growth conditions. We studied the surface morphology using interference microscopy and profilometry (such as Dektak). In some cases, detailed analyses were done using SEM imaging and EBSD techniques to study the orientation of the defects. Detailed analyses of defects were done by chemical etching using Sopori etch. Defect imaging of the whole wafer (125mm x 125mm) was done using a reflectometer imaging technique described in ref [2]. Defects within the wafer thickness were exposed by cross-sectioning the wafer, followed by defect etching. Minority carrier lifetime is measured on these wafers using Iodine ethanol passivation for estimation of the bulk material quality and the cell response was analyzed using LBIC imaging.

Surface morphology

It is well known that epitaxial growth is very sensitive to substrate cleanliness, particulate matter deposited during the wafer growth, gas flow conditions, and the epitaxial growth rate. Some growth conditions show a variety of surface morphologies that are reminiscent of the Si-bearing gas flow patterns, which get etched off during the wafer processing. Figure 2(a) is a reflectance map generated in diffused mode (measuring only the scattered light) showing the top surface characteristics. It may be noted that the diffuse reflectance of the wafer is quite low, indicating high surface planarity. All the high reflectance dots in the image are the pyramids formed by the stacking faults (SFs). The Gas flow pattern is also visible in the picture. Figure 2(b) is Dektak scan across the wafer surface in the middle of the wafer showing the typical roughness (p-t-v, peak to valley) of an epi wafer (~0.2μm). Figure 3 shows microscopic images of the epi wafer surface showing mounds, saucers, and pyramid. Figure 4(a) shows a Dektak scan across a pyramid on the wafer surface (pyramid height ~6μm), and fig. 4(b) shows an interference microscope image a pyramid at a different location.

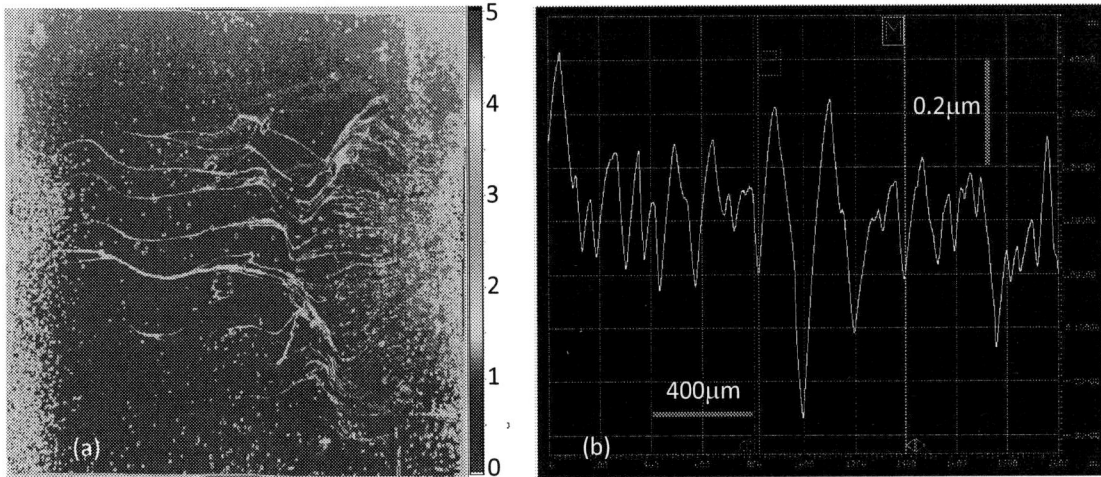

Figure 2. (a) A reflectance map of a wafer grown under conditions that cause flow patterns and particulate embedding in the epi layer (b) A Dektak trace showing typical roughness of an epi wafer front surface. The average p-t-v variations are about 0.3 μm.

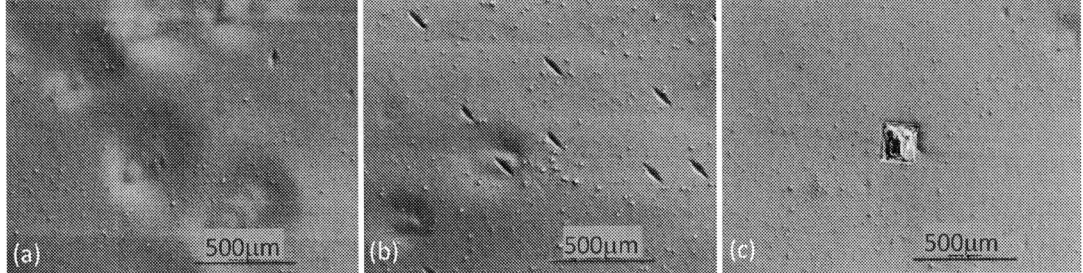

Figure 3. Optical microscope images of the wafer surface that has gas flow pattern, it consists of: (a) mounds, (b) shallow saucers, and (c) pyramids.

Defect etching results

Defect etching was done to delineate the crystallographic defects and to determine to what extent the surface morphology is related to the bulk defects such as dislocations and stacking faults. It was found that, isolated regions of dislocations do not occur very often. When formed, they appear as dislocation loops, indicating that generation caused by either a precipitate or a particulate matter. Figure 5 (a) shows an optical microscope image of a defect etched epi wafer showing multiple dislocations (in the form of etch pits) forming a local network of dislocations. The average dislocation density is $<10^3$ cm^{-2}.

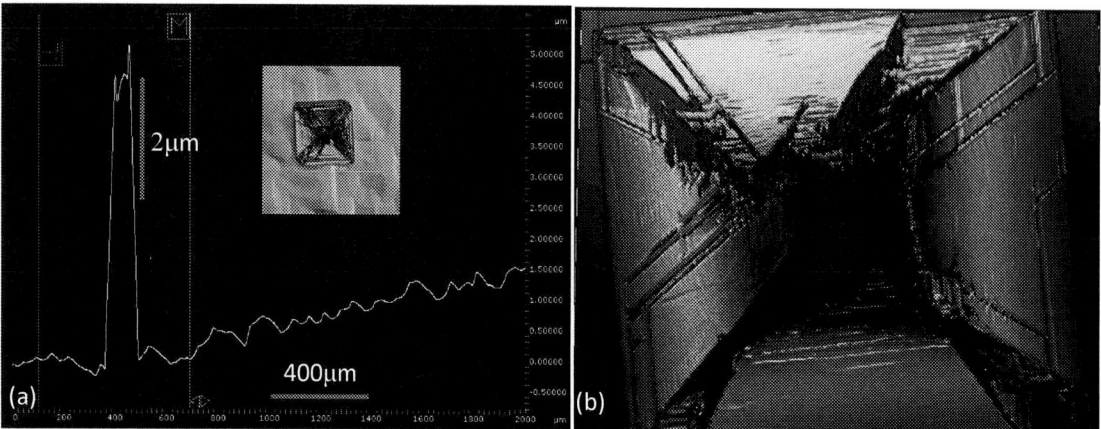

Figure 4. (a) Dektak scan of a pyramid on wafer surface, and (b) Interference microscope image of the pyramid. Inset of figure (a) is the optical microscope image of the actual pyramid across which the Dektak scan was run.

Stalking faults are known to be the main defect in epitaxial growth [3]. They may occur in many different configurations. Figure 5 also shows etch patterns for various types staking faults observed in the epi wafers. Figure 5(b) shows isolated SFs along <110> direction and can be of different length. Figure 5(c) shows intersecting multiple stacking SFs that form squares. And Figure 5(d) depicts nested SFs (Intersecting SFs formed with in other intersecting SFs). The various types of SFs can be highly localized in few regions of the wafer. It is known that in thin epitaxial layers, SFs mostly originate from the interface and grow through the film [4].

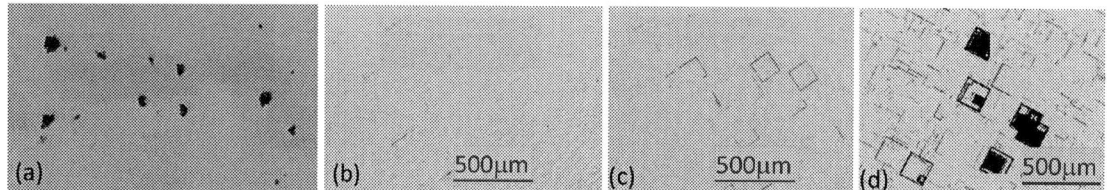

Figure 5. Microscope images of (a) local dislocation networks (b) isolated SFs (c) intersecting SFs, and (d) nested SFs.

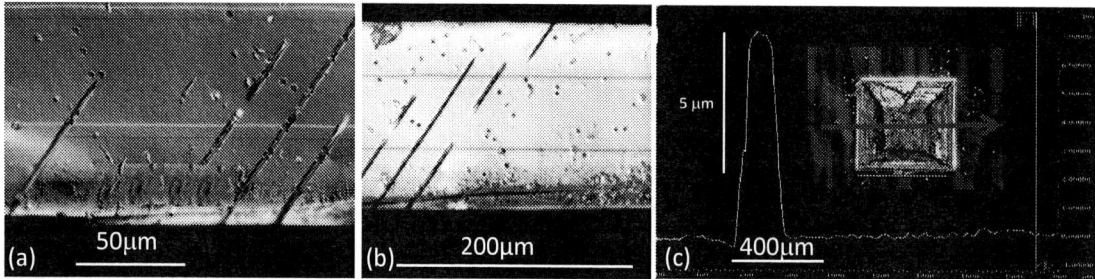

Figure 6. (a), (b) microscope images of an epi wafer cross section after defect etch, and (c) Dektak trace over a defect etched pyramid surface, Inset show the actual pyramid decorated with dislocations.

Figure 6 shows microscopic images defect etched cross sections of a wafer showing that SFs can originate either at the substrate-epi interface as well as within the wafer thickness. Stacking faults originating within the epi have not been reported previously.

Large, nested SFs can protrude several micrometers out of the wafer surface forming pyramid and punch out dislocations. Figure 6 also shows a Dektak trace taken along the arrow on a pyramid, showing the height of the pyramid to be about 8 µm. The inset is a defect etched image showing the pyramid consists of a fine structure (etched lines) and punched out dislocations. Clearly, the pyramid formation also generates a local stress, which is responsible for localized dislocation generation.

Dislocation generation and the fine structure of the pyramid can be seen from defect etched cross sectioned view shown in Fig 7 (a) and (b). We have observed, contrary to the common belief that pyramids can originate within the epi. We believe that this type of pyramids are started by an unusual SF formation where (i) the initial SF occurs parallel to the (100) surface. We have called this to be the primary fault, then (ii) the pyramid grows by the formation of the secondary SFs parallel to the (111) planes.

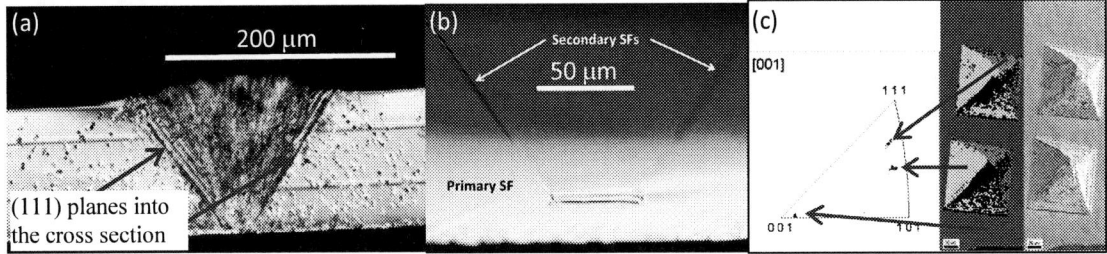

Figure 7. Microscope images of cross sectioned pyramids (a) starting from epi-substrate interface and (b) starting within the bulk of the epi wafer, and (c) pole figure, EBSD and SEM images (showing coherent twinning) of two adjacent pyramids.

Our EBSD analyses have determined that the "fine structure" of the pyramid consists of coherent twins. Figure 7 (c) shows pole figures of various planes of two adjacent pyramids. Because of the angle of the pyramid faces, the poles do not correspond to the (111) planes. Multiple twinning can also be seen from SEM images of these pyramids. One can associate pyramids to be some form of local dendritic growth, where the growth rate is higher than the main wafer.

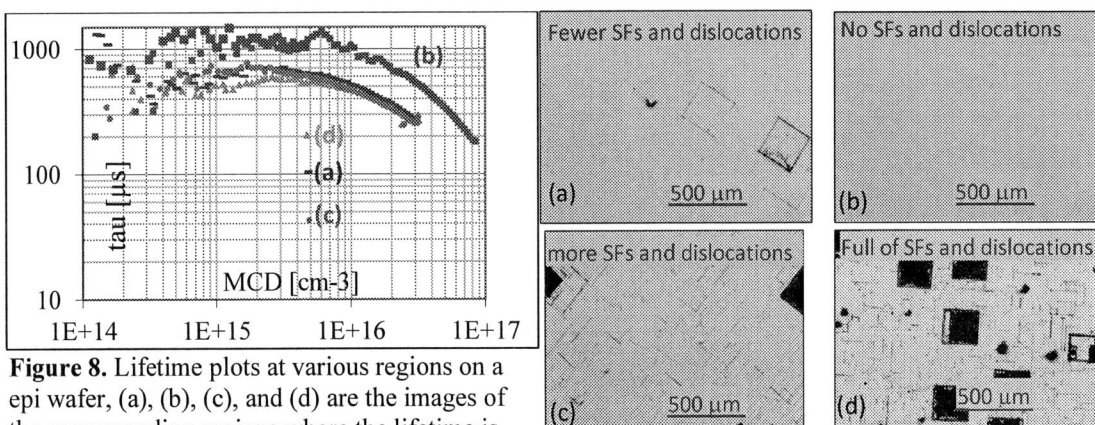

Figure 8. Lifetime plots at various regions on a epi wafer, (a), (b), (c), and (d) are the images of the corresponding regions where the lifetime is measured.

Influence of defects on the minority carrier lifetime and the cell performance

We measured minority carrier lifetimes in various regions decorated with defects. Figure 8 is a typical result of these measurements. We have shown lifetime as a function of injection level measured in regions which have typical defects as seen by the microscope images of the defect etched wafer. It is seen that highest lifetime of about 1.2 ms occurs in defect-free regions. Lowest lifetime occurs in regions with excessive pyramid formation. A similar behavior is seen in the final cells. Figure 9 shows LBIC response of a solar cell taken at a wavelength of 0.98 μm. The response from majority of the cell is about 0.6 mA/mW, but there are regions of lower response (0.4 to 0.5 mA/mW), which are affected by the pyramid formation.

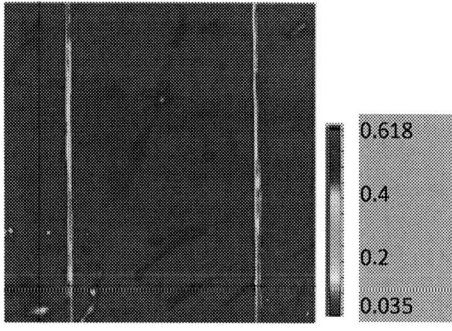

Figure 9. LBIC response of a solar cell fabricated on a p-type, epitaxially grown wafer. This cell was fabricated on a p-type epi wafer using standard cell fabrication processes. The efficiency of the cell was 17.5% under AM1.5 illumination. It should be pointed out that in absence of the pyramids, this cell would be about 20% efficient.

CONCLUSION/DISCUSSION

Our study has shown that defect generation mechanisms are of two types: (i) Type A- interface defects that originate/nucleate from factors such as surface cleanliness, quality of the porous Si at the surface, and factors related to the nucleation kinetics at the initial growth, and (ii) Type B- propagation of the interface defects and generation of bulk defects through thermal stress. Type A defects are predominantly stacking faults (SF), while type B are primarily dislocations. Analyses of the solar cells have revealed an interesting behavior — in spite of the fact that SF density is low, they can have a controlling effect on the solar cell performance. As pointed out earlier in this work, the epitaxial wafers investigated in this work were grown from a prototype reactor. Manual operation of this system introduces high levels of contamination and correspondingly crystal defects in many forms have been generated in the wafers. However, significantly reduced defects and impurities have been noticed in the epitaxial wafers which were grown in a production type reactor with system level automation. Higher lifetime for both N-Type and P-type epitaxial wafers have been measured with these wafers, and solar cell efficiency >20% has been achieved [5].

REFERENCES

1. Duerinckx. F. et al, Progress in Photovoltaics, **13**, (2005), pp673.
2. B. Sopori. Et al, IEEE PVSC, (2010), pp2238.
3. D. Pomerantz, Journal of Applied Physics, **38**, (1967), pp5020.
4. J Washburn et al, Applied phyicss letters, **3**, (1963), pp44.
5. Ruiying Hao, T.S. Ravi, V. Siva, Jean Vatus, Dan Miller, Joel Custodio, Ken Moyers, Chia-Wei Chen, Ajay Upadhyaya, Ajeet Rohatgi, Proc. IEEE PVSC, 2014, Denver, CO, to be published.

Mater. Res. Soc. Symp. Proc. Vol. 1666 © 2014 Materials Research Society
DOI: 10.1557/opl.2014.920

Crystallization of Amorphous Silicon Thin Films by Microwave Heating

Tomohiko Nakamura[1], Shinya Yoshidomi[1], Masahiko Hasumi[1], Toshiyuki Sameshima[1], and Tomohisa Mizuno[2]
[1]Tokyo University of Agriculture and Technology, Tokyo, 184-8588 Japan
[2]Kanagawa University, Kanagawa, 259-1293 Japan

ABSTRACT

We report crystallization of amorphous silicon (a-Si) thin films and improvement of thin film transistors (TFTs) characteristics using 2.45 GHz microwave heating assisted with carbon powders. Undoped 50-nm-thick a-Si films were formed on quartz substrates and heated by microwave irradiation for 2, 3, and 4 min. Raman scattering spectra revealed that the crystalline volume ratio increased to 0.42 for the 4-min heated sample. The dark and photo electrical conductivities measured by Air mass 1.5 at 100 mW/cm^2 were 2.6×10^{-6} and 5.2×10^{-6} S/cm in the case of 4-min microwave heating followed by 1.3×10^6-Pa-H_2O vapor heat treatment at 260°C for 3 h. N channel polycrystalline silicon TFTs characteristics were improved by the combination of microwave heating with high-pressure H_2O vapor heat treatment. The threshold voltage decreased from 5.3 to 4.2 V and the effective carrier mobility increased from 18 to 25 cm^2/Vs.

INTRODUCTION

Crystallization of amorphous silicon thin films is an important processing technology for fabricating the thin film transistors (TFTs) with a high carrier mobility and low threshold voltage, which can be applied to switching and driving circuits in flat panel displays [1,2]. Laser crystallization using XeCl excimer laser has been widely used for crystallization of amorphous silicon (a-Si) thin films [3]. However, laser crystallization requires a complicated equipment and high operation cost. These problems give us a motivation of development of simple crystallization technique at a low cost. In this paper, we report a crystallization method using microwave heating with carbon powders. We report crystallization condition of a-Si films and their crystallographic and electrical properties. In addition, we discuss improvement in TFT characteristics by the present microwave heating.

EXPERIMENT

50-nm-thick a-Si films were formed on quartz substrates with a diameter of 4 inch using low pressure chemical vapor deposition (LPCVD). The samples were divided into 4 quarter pieces. The samples were placed in a quartz vessel with an internal diameter of 2.4 inches. Carbon powders at 12 g were also put in the vessel to completely cover the sample, as shown in Fig. 1. The vessel was placed in a 2.45 GHz commercial microwave oven. Microwave irradiation was operated at 1000 W for 2, 3, and 4 min. Carbon powders effectively absorbed microwave power. They began to emit blight red light at 2 min and emitted blight red orange light at 4 min. According to the blackbody radiation theory, carbon powders were heated to about 1000°C for 4 min microwave irradiation. The samples were then heated with 1.3×10^6 Pa H_2O vapor at 260°C for 3 h to reduce defect states of silicon films [4-6]. The samples were heated by heat conduction from carbon powders.

Raman scattering spectra were measured from 400 to 600 cm⁻¹ using a 514.15 nm probe laser beam and analyzed using a numerical calculation program to estimate the crystalline volume ratio. For measurement of the electrical conductivity, aluminum electrodes were formed on the sample surface by vacuum evaporation. The size of Al electrodes were 1 mm² and the gap length between electrodes was 1.5 mm. The electrical conductivity was measured in the dark and under Air mass 1.5 light illumination at 100 mW/cm².

Microwave heating was also applied to improve TFT characteristics. N channel polycrystalline TFTs were prepared on a quartz substrate using plasma enhanced CVD of 50-nm-thick silicon and 100-nm-SiO₂ gate insulator layers and laser crystallization. The channel width and length were 75 and 25 μm, respectively. The sample was cut to pieces with a size of 5x5 mm². TFT were heated in carbon powders by microwave heating for 1.5 min followed by high pressure H₂O vapor heat treatment with 1.3x10⁶ Pa at 260°C for 3 h. Transfer and output characteristics were measured using agilent 4156C.

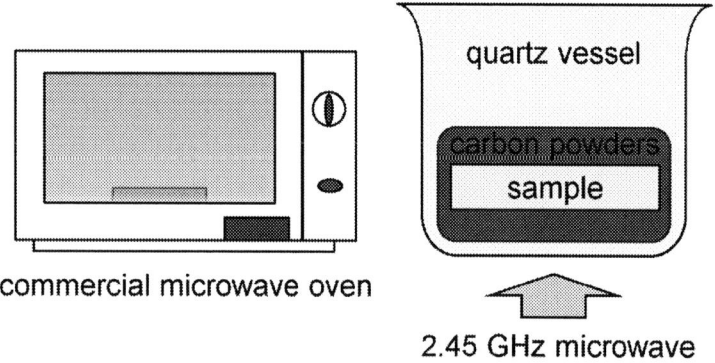

Figure 1. Schematic cross section of experimental equipment and microwave irradiation image for heating the samples.

DISCUSSION

Figure 2 shows Raman scattering spectra of the 50-nm-thick amorphous silicon films with different microwave heating durations. The initial sample had a broad peak at about 470 cm⁻¹ with a full width at half maximum (FWHM) of 88.7 cm⁻¹, which shows typical amorphous lattice vibration mode. The crystalline volume ratio was 0. For 2-min heated sample, the edge region about 1 cm was crystallized and showed a sharp peak at 515.5 cm⁻¹ with a FWHM of 11.4 cm⁻¹ and the crystalline volume ratio of 0.37. On the other hand, the internal region was still amorphous, whose Raman spectra had a broad peak at about 470 cm⁻¹ with a FWHM of 100 cm⁻¹. In the cases of 3 and 4 min microwave irradiations, the samples were entirely crystallized over the whole region with sharp Raman scattering peaks, as shown in Fig. 2. Raman scattering spectra at the middle point gave the peak wavenumber, FWHM and crystalline volume ratio of 515.1 cm⁻¹, 11.9 cm⁻¹, and 0.35 for 3-min heated sample, and of 515.3 cm⁻¹, 11.5 cm⁻¹, and 0.42 for 4-min heated sample, respectively. These results demonstrated that microwave heating with carbon powders easily and rapidly crystallized a-Si films. The crystalline volume ratio slightly increased as the heating duration increased from 2 to 4 min. We interpret that the crystallization initialized in the solid state under the high density of crystalline nucleations at a low temperature in the initial stage of the microwave heating, and that crystalline grains grew and the amorphous region gradually reduced with increasing temperature as the heating duration increased. 4-min microwave irradiation probably heated the silicon film at the highest temperature and crystallized with the highest crystalline volume ratio.

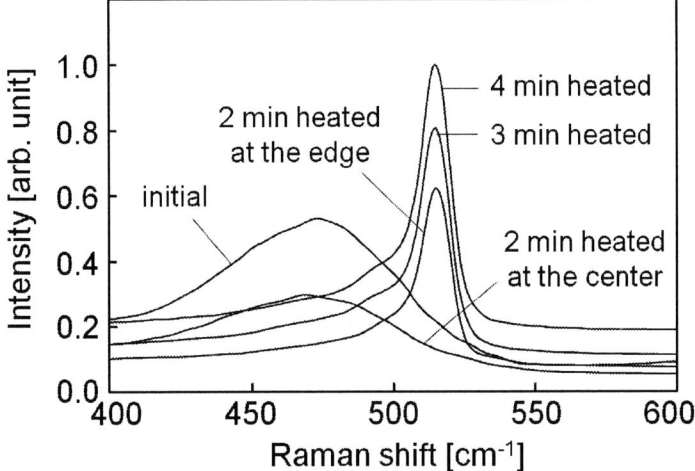

Figure 2. Raman scattering spectra of the 50-nm-thick amorphous silicon films with different microwave heating duration.

Figure 3 shows dark (solid circles) and photo (open circles) electrical conductivities as a function of duration of microwave irradiation for initial a-Si and as-microwave heated samples. Figure 3 also shows dark (solid triangles) and photo (open triangles) electrical conductivities as a function of duration of microwave irradiation for initial and high pressure H_2O vapor annealed samples. In the case of 2-min microwave irradiation, electrical conductivities at central and edge regions were presented. For samples crystallized by microwave heating, the dark conductivities ranged from 6.2×10^{-7} to 1.4×10^{-6} S/cm and the photo conductivity ranged from 1.4×10^{-6} to 2.7×10^{-6} S/cm, which was much lower than that of initial a-Si. The decrease in photo conductivity indicates decrease in the absorption coefficient in the visible range owing to change from the direct to indirect transition due to crystallization. On the other hand, H_2O vapor heat treatment increased the dark conductivity from 2.6×10^{-6} to 1.0×10^{-5} S/cm and increased photoconductivity from 5.2×10^{-6} to 2.1×10^{-5} S/cm among from 2 to 4 min crystallized sample. We interpret that the surface and grain boundaries were passivated by high-pressure H_2O vapor heat treatment and recombination probability of photo-induced carriers was reduced. The central region of 2-min heated sample with amorphous state, as shown in Fig.2, had very low dark and photo conductivities. They indicate that the region had a high band gap with a low carrier density and had a high density of defect states. Although the photo conductivity was increased by the high pressure H_2O vapor heat treatment, it was still much lower than that initial a-Si. Defect reduction was not probably sufficient.

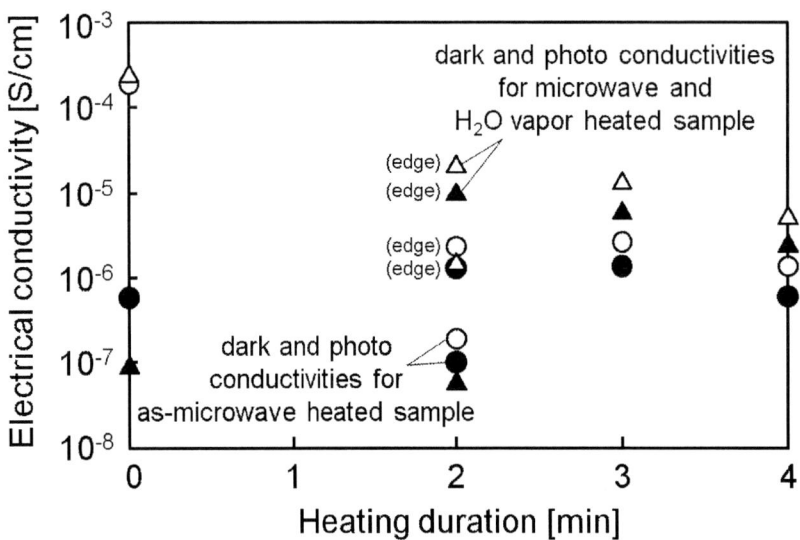

Figure 3. Dark and photo-electrical conductivities for as-microwave heated samples and for microwave heated and H_2O vapor heated samples at central region and edge region (edge).

Figure 4 shows transfer characteristics at a drain voltage of 0.1 V for initial, as-1.5 min microwave heated, and microwave heated and then H_2O vapor heated samples. Initial sample showed gradual increase in the drain current by increasing the gate bias voltage. The analysis of the transfer characteristics resulted in the threshold voltage V_t of 5.3 V and the effective carrier mobility μ_{eff} of 18 cm^2/Vs. On the other hand, 1.5 min microwave heating markedly decreased the drain current in the positive gate bias voltage condition, as shown in Fig. 4. The V_t increased to 13 V and μ_{eff} decreased to 16 cm^2/Vs, while no damage was observed by optical microscope observation. The drain current was sharply improved as the gate voltage increased when the sample was heated with 1.3×10^6 Pa H_2O vapor at 260°C for 3 h. The V_t decreased to 4.2 V and μ_{eff} increased to 25 cm^2/Vs. We interpret that 1.5 min microwave heating improved crystalline state especially in regions around grain boundaries but caused the increase of point defect type trap states because of rapid heating with high temperature. Subsequent H_2O vapor heat treatment probably reduced the density of defect states generated by microwave heating. We believe therefore that the combination of microwave heating with H_2O vapor heat treatment has a possibility of improvement in TFT characteristics.

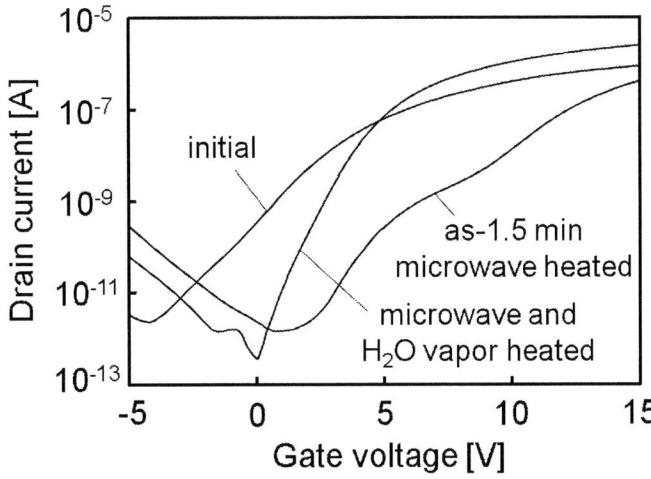

Figure 4. Transfer characteristics for initial, as-1.5 min microwave heated, and microwave and H_2O vapor heated samples.

Figure 5 shows output characteristics for initial (a), as-1.5 min microwave heated (b), and microwave and H_2O vapor heated samples (c). The initial TFT had typical linearly ohmic characteristic in a low drain voltage region and typical pinch off characteristic in a high drain voltage region. 1.5 min microwave heating markedly decreased the drain currents. On the other hand, high-pressure H_2O vapor heat treatment increased the drain currents and recovered typical linearly ohmic characteristic in a low drain voltage region and typical pinch off characteristic in a high drain voltage region. These results indicate that the TFT structure well maintained during microwave and H_2O vapor heating, and that the present heating method realized crystalline state with a low density of defect states to allow TFT operation with low voltages.

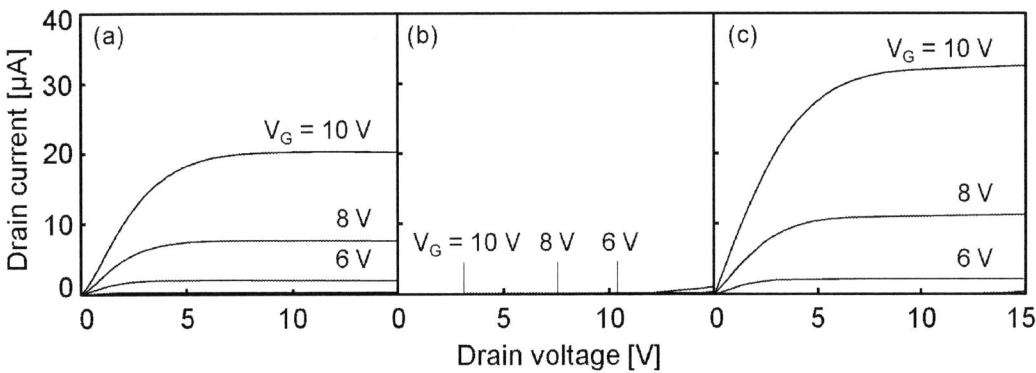

Figure 5. Output characteristics for initial (a), as-1.5 min microwave heated (b), and microwave and H_2O vapor heated samples (c).

CONCLUSIONS

We reported crystallization of a-Si thin films and improvement of TFTs characteristics using 2.45 GHz microwave heating assisted with carbon powders. Undoped 50-nm-thick a-Si films were formed on quartz substrates and heated by microwave irradiation for 2, 3, and 4 min using a commercial microwave oven. Carbon powers effectively absorbed microwave energy and heated to high temperature up to 1000°C. Silicon samples were heated via heat conduction from carbon. Raman scattering spectra revealed that a-Si films were rapidly crystallized by microwave heating. The crystalline volume ratio increased to 0.42 as heating duration increased to 4 min. The dark and photo electrical conductivities measured by Air mass 1.5 at 100 mW/cm^2 were 2.6×10^{-6} and 5.2×10^{-6} S/cm in the case of 4-min microwave heating followed by 1.3×10^6-Pa-H_2O vapor heat treatment at 260°C for 3 h, although photo electrical conductivity was low of 1.4×10^{-6} S/cm just after microwave heating probably because substantial light trapping defects generated during microwave heating. N channel polycrystalline silicon TFTs characteristics were improved by the combination of 1.5-min microwave heating with high-pressure H_2O vapor heat treatment. The threshold voltage decreased from 5.3 to 4.2 V and the effective carrier mobility increased from 18 to 25 cm^2/Vs.

ACKNOWLEDGEMENTS

This work was partly supported by Grant-in-Aid for Scientific Research C (#25420282 and #23560360) from the Ministry of Education, Culture, Sports, Science and Technology of Japan and Sameken Co., Ltd.

REFERENCES

1. T. Serikawa, S. Shirai, A. Okamoto and S. Suyama, " Electrical Characteristics of High-Mobility Fine-Grain Poly-Si TFTs from Laser Irradiated Sputter-Deposited Si Film", Jpn. J. Appl. Phys. 28 (1989) 1871
2. S. Uchikoga and N. Ibaraki, " Low temperature poly-Si TFT-LCD by excimer laser anneal", Thin Solid Films 383 (2001) 19.
3. T. Sameshima, M. Hara and S. Usui, "XeCl Excimer Laser Annealing Used to Fabricate Poly-Si TFTs", Jpn. J. Appl. Phys. 28 (1989) 1789-1793.
4. T. Sameshima and M.Satoh, "Improvement of SiO_2 Properties by Heating Treatment in High Pressure H_2O Vapor", Jpn. J. Appl. Phys. 36 (1997) L687-L689.
5. T. Sameshima, M. Satoh, K. Sakamoto, A. Hisamatsu, K. Ozaki and K. Saitoh, "Heat Treatment of Amorphous and Polycrystalline Silicon Thin Film with H_2O Vapor", Jpn. J. Appl. Phys. 37 (1998) L112-L114.
6. T. Sameshima, M. Satoh, K. Sakamoto, K. Ozaki and K. Saitoh, "Heat Treatment of Amorphous and Polycrystalline Silicon Thin Films with High-Pressure H_2O Vapor", Jpn. J. Appl. Phys. 37 (1998) 4254-4257.

AUTHOR INDEX

Ariosa, D. 78

Badan, J. 78

Bakker, N. 42

Basnyat, P. 109

Beyer, W. 36, 85

Bidiville, A. 7

Cormier, D. 91

Custodio, J. 48

Dalal, V. 97

Dalchiele, E. 78

Devayajanam, S. 109

Dubey, M. 1

Dutta, P. 103

Fan, Q. 1

Fernandes, M. 59

Furukawa, J. 18

Gallagher, J. 24

Hao, R. 48, 109

Hasumi, M. 18, 115

Hilgers, W. 36

Hirschman, K. 91

Jiang, C.-S. 1

Kondo, M. 7

Konduri, S. 97

Kouvetakis, J. 24

Kuang, Y. 42

Leinen, D. 78

Lennartz, D. 36

Louro, P. 65, 71

Luekermann, F. 85

Maier, F. 36

Manley, R. 91

Marotti, R. 78

Martin, F. 78

Matsui, T. 7

Matsumoto, M. 7

Menendez, J. 24

Miller, D. 48

Mizuno, T. 18, 115

Moutinho, H. 109

Moyers, K. 48

Mudgal, T. 91

Mulder, W. 97

Nag, S. 109

Nakamura, T. 115

Nickel, N. 36

Node, T. 18

Pennartz, F. 36

Pfeiffer, W. 85

Prunici, P. 36, 85

Ramos-Barrado, J. 78

Ravi, T. 48, 109

Reedy, R. 109

Reepmeyer, C. 91

Rodrigues, I. 65, 71

Rumyantsev, S. 103

Sai, H. 7

Saito, K. 7

Sameshima, T. 18, 115

Saxena, T. 103

Sazonov, A. 59

Schropp, R. 42

Shigeno, S. 18

Shur, M. 103

Silva, V. 65, 71

Siva, V. 48

Soleymanzadeh, B. 85

Sopori, B. 109

AUTHOR INDEX

Stevenson, D....................................... 1

Stiebig, H.. 85

Suezaki, T....................................... 7

Toner, W. 1

VanSant, K..................................... 109

Vatus, J.................................... 48, 109

Veldhuizen, L.................................. 42

Vieira, M. 59, 65, 71

Vygranenko, Y................................ 59

Werf, C... 42

Yan, B... 1

Yoshida, I...................................... 7

Yoshidomi, S............................ 18, 115

Yun, S... 42

Cambridge University Press
32 Avenue of the Americas, New York, NY 10013-2473, USA

Materials Research Society
506 Keystone Drive, Warrendale, PA 15086

ISBN 978-1-5108-0527-9